T0258636

IET HISTORY OF TECHNOLOGY SERIES 12

Series Editor: Dr B. Bowers

A History of the World Semi-Conductor Industry

Other volumes in this series:

A History of the World Semi-Conductor Industry

P.R. Morris

The Institution of Engineering and Technology in association with The Science Museum, London

Published by The Institution of Engineering and Technology, London, United Kingdom

First edition © 1990 Peter Peregrinus Ltd
Reprint with new cover © 2008 The Institution of Engineering and Technology

First published 1990
Reprinted 2008

The Institution of Engineering and Technology
Michael Faraday House
Six Hills Way, Stevenage
Herts, SG1 2AY, United Kingdom

www.theiet.org

British Library Cataloguing in Publication Data
Morris, P.R.
  History of the world semiconductor industry
  (History of technology series no 12)
  1. Semiconductor industries, history
  I. Title II. Series
  338.47621381520904

ISBN (10 digit) 0 86341 227 0
ISBN (13 digit) 978-0-86341-227-1

First printed in the UK by Short Run Press Ltd, Exeter
Reprinted in the UK by Lightning Source UK Ltd, Milton Keynes

Contents

Preface

This book, entitled 'A history of the world-wide semiconductor industry', is a general historical survey covering events leading to the establishment of the semiconductor industry and its subsequent development up to about the middle of the 1980s. Although a description of the technical developments of semiconductor devices during this period forms an essential part of the work, treatment in this respect has been largely of a descriptive nature, with the minimum of technical detail.

The origins of semiconductor development are largely bound up with that of the thermionic valve industry, and this aspect has been taken into account by including a brief description in chapter 2 entitled 'The development of the thermionic valve'.

The growth of the semiconductor industry has been considered within four basic geographical divisions, namely the United States, Western Europe, Eastern Europe and Japan. The decision to adopt this approach was more than a matter of convenience, since the national semiconductor industries included within these areas appear to share certain common problems, and have evolved in a similar manner. Nevertheless, in view of their importance, the international character of the industry and the manner of diffusion of technical information and manufacturing processes have also received attention.

The final chapter opens with a discussion of the nature of semiconductor technology and its relationship with science, and this is followed by a review of the factors influencing the growth of the industry within the major geographical areas outlined above. Both the importance of Europe retaining a strong manufacturing base in this field and the prerequisites for this are emphasised. The chapter concludes with the recommendation that in view of the all-pervading nature of the rapid changes in society resulting from the development of semiconductor technology, an urgent need exists to understand the nature of these changes.

A significant proportion of this book rests upon information obtained through my experience as a semiconductor device engineer within the industry and numerous conversations I have had over the years with my colleagues. I am also indebted to those who have helped me directly by granting interviews and supplying information. In particular, I would like to thank Professor R.A. Buchanan, Director of the Centre for the History of Technology, Science and Society, University of Bath, and Mr B.N. Cotton, Dean, Southampton Institute of Higher Education, for their advice and encouragement, also Dr D. Parkinson, Ex. Head of Physics, TSRE, Malvern, for his critical appraisal, and Mr J. Bolt, Senior Lecturer, Southampton Institute of Higher Education, for his careful checking of technical data and some helpful suggestions. Finally, a special word of thanks to my daughter Siani who, armed with a word processor and a keen eye for detail, has considerably simplified my task.

P. R. Morris

Chapter 1
Introduction

This book reviews the growth of semiconductor microelectronics within the principal nation-states currently possessing such an industry, and also includes an account of relevant inventions and discoveries assisting that growth.

Important aspects which are considered include the relationship between science and semiconductor technology, the process of diffusion of technology and the influence of economic and military requirements upon the manner of industrial growth. As these various topics have an interrelated aspect, they have largely been dealt with within the context in which they have arisen, rather than in isolation.

Perhaps the most important external factor influencing the early development of semiconductor devices was the thermionic valve industry. Consequently, discussion of the growth of the semiconductor industry must take into account the reasons for the initial success and eventual decline of its thermionic rival. Therefore, before embarking upon an account of the development of semiconductor devices, Chapter 2 offers a brief historical review of that of the thermionic valve from the time of its invention until the end of the Second World War.

Since the origins of present semiconductor technology are rooted in nineteenth century experimental science, this period has been made a starting point for a review of semiconductor device development and forms the subject of Chapter 3. Specifically, this chapter looks at research in the early semiconductor field and the development of semiconductor devices from their beginning up to the period immediately before the invention of the transistor. The theoretical framework within which this work took place is also briefly discussed in a non-analytical manner.

Chapter 4 is concerned with the development of solid state devices from the time of the invention of the transistor until about the middle of the present decade. Firstly, the major types of pre-planar devices are considered, this being followed by an account of the planar transistor and its logical successor, the integrated circuit. Later developments such as the microprocessor, the semi-custom circuit and gate array are also reviewed. An account of the history of the field-effect transistor has been included in this chapter because of its important role in the development of integrated circuitry and a discussion of the influence of large-scale integration upon systems design considers technical problems raised by these developments. In addition, in order to illustrate the diversity of device types, mention has been made of the small-signal diode, the power rectifier and the thyristor.

Chapter 5 describes the major technical processes used in semiconductor device fabrication up to the early 1980s. The adaptation of these key processes to semiconductor manufacture on a mass-production basis has largely resulted in the rapid improvements in device technology which are a characteristic of this particular industry.

A consideration of the development of semiconductor device manufacture on a world-wide scale shows that it may be conveniently divided into a fairly small number of geographical centres, each of which may be considered independently. Adopting this approach, Chapter 6 deals with the development of the semiconductor manufacturing industry in the United States. This country is of particular importance, not only as the birthplace of the transistor, but also because at least until the middle of the 1980s it almost completely dominated the industry on an international basis. The only effective challenge to this situation has been made by Japan and Chapter 7, in reviewing the history of the semiconductor industry in that country, also considers the nature of this challenge and the reasons for its success. Another recent event of importance has been the rise of an indigenous South Korean semiconductor industry and this development is also considered within this chapter.

The third largest geographical group engaged in solid-state device production is Europe. Chapter 8 begins by reviewing the general characteristics of the transistor industry within Western Europe and then goes on to consider in turn the development of semiconductor manufacturing in the United Kingdom, France, Italy and Germany. This chapter also includes a brief assessment of the Soviet and East-European semiconductor industries.

The final chapter in this book attempts to bring together the material and ideas previously expressed, firstly in order to review the growth of the industry in terms of its relationship to science and technology, and then to consider reasons for its relative performance within the three main geographical areas outlined in Chapters 6, 7 and 8. It concludes with an assessment of the prospects of the European semiconductor industry, and advocates the need for a critical evaluation of the various problems raised by its rapid emergence.

In addition to the above general outline, it may be helpful within this introduction to define the major boundaries within which the work was undertaken. These are as follows:

Countries in which semiconductor assembly is carried out under foreign control, or in which this activity forms the major part of their device industry are not reviewed separately in this study, since this activity is considered as being an overseas operation under the control of the parent company. Consequently, developments in such countries as Sri Lanka, Singapore, the Philippines and Malaysia have been omitted, since they fall into this category. South Korea is an exception, because that nation, although originally only possessing foreign assembly plant, has more recently built up an indigenous semiconductor industry of its own and is therefore reviewed in Chapter 7.

In view of the great variety of semiconductor types, it was felt necessary to concentrate entirely upon the main materials used in device production, namely silicon and germanium, ignoring such semiconductor compounds as gallium arsenide and also those used in the manufacture of photocells, for instance lead sulphide, cadmium sulphide, cadmium mercury telluride etc. Although of importance in specific applications, devices using such semiconductor compounds may be regarded as being outside the mainstream of device manufacture and except in specific applications within the military field have had little bearing so far upon the general development of the semiconductor industry. Nevertheless, in the interests of completeness, an appendix briefly outlining the characteristics of gallium arsenide and its applications has been included.

In reviewing the history of the various national semiconductor manufacturing industries, a balance of emphasis has been attempted based upon their relative importance as producers of devices. However, an exception has been made in the case of the United Kingdom, its industry being treated in somewhat more detail. This was done in order to illustrate in rather more depth the general problems currently facing European semiconductor manufacturers within the international market.

Treatment of the Soviet semiconductor industry and its Eastern bloc allies has been rather under-represented. This is largely because the manufacture of solid state devices in the Soviet Union is under the control of the Ministry of the Electronics Industry (Minelektronprom) and is therefore part of the Soviet defence industry. Consequently, information on the history of semiconductor device technology within that country has been difficult to obtain.

Chapter 2
Development of the thermionic valve

The origins of semiconductor devices were inextricably bound up with those of the thermionic valve. Consequently, any account of the semiconductor industry must necessarily be incomplete without some consideration of how the vacuum tube industry grew, flourished, and then fell into relative decline, the latter largely due to the success of its solid-state rival. As emphasis has been placed on those fields where the influence of the semiconductor impinged most greatly upon that of the vacuum tube, this account must necessarily be incomplete.

2.1 Development of the diode and triode

The early beginnings of the thermionic industry can be traced back to the 1880s and the incandescent lamp industry. Indeed, the electric lighting business formed the basis from which, by way of research, the thermionic valve industry developed. Experiments by Edison at this time showed that current would flow if a wire inserted within a light bulb was connected to a positive potential. This effect (known as the Edison effect) was described in 1880[1,2]. However, Edison was strongly opposed to the use of alternating current and did not investigate the possibility of rectification. A patent application was made by Professor J.A. Fleming describing a two-electrode valve for the rectification of high frequency alternating currents.[3] However, earlier in the same year A. Wehnalt, working in Germany, had already patented a thermionic diode which he clearly recognised as a rectifier (January 1904). But Stokes points out that:

> 'Fleming was able to patent his valve in Germany so the German Patent Office apparently found nothing which could have been regarded as a prior claim.'[4]

Soon afterwards (25th October 1906) Lee de Forest, working in the United States, applied for a patent for a triode valve—a three electrode thermionic device capable of amplification of electric current and this patent was granted in February 1908.[5] Although de Forest recognised the increased sensitivity of the triode as a detector compared with the vacuum diode, it does not seem to have been recognised that the triode could amplify. This discovery seems to have been made independently by Von Lieben in Germany and Edwin Armstrong in America in 1911. The invention of this device led to bitterness and litigation involving Fleming and de Forest, which lasted until 1943 when the United States Supreme Court finally decided that the original Fleming patent was invalid insofar as it did not cover the triode amplifier.

The early triode valve was not widely used and the theory of its operation was not understood. W.A. Atherton writes 'like the early diode, the early triode was not a great commercial success'.[6] However, in 1912 de Forest demonstrated a multistage amplifier using triodes to the American Telegraph and Telephone Company (AT & T) where the idea was rapidly improved. In 1914, that company bought the radio receiver rights for the triode for a sum of $90,000. Atherton comments:

> 'It was thanks to the work of large industrial laboratories, especially those of A.T. & T. and G.E. (General Electric Company) that the crude triode was transformed into a reliable and efficient device.'[7]

Advances in valve circuitry began to be made immediately before the outbreak of the First World War. For instance, C.S. Franklin (UK) discovered the phenomenon of feedback for high-frequency amplification in 1912[8] and in the following year H.J. Round (UK) described the use of the triode as an oscillator[9], although there were several other claimants to these inventions, including de Forest.

Early valves were known as 'soft' valves. They suffered from positive ion bombardment of the cathode, resulting in a short life. Improvements during the First World War in vacuum technology led to the construction of 'hard' valves in which the envelope was highly evacuated, this leading to much less contamination of the cathode by ionic bombardment and consequently a longer life. Sir E. Appleton writes:

> 'The war really killed off the 'soft' valve, since in military operations, tubes were required which possessed the characteristics of consistency, reproducibility and reasonably long life.'[10]

He also states:

> 'From the scientific point of view, the chief result of the First World War was that people began to understand how a valve worked and also how its performance as a detector, amplifier and oscillation generator was determined by its internal physical dimensions and structure.'

Certainly the war led to great improvements in thermionic valve technology, emphasis being placed upon performance, reliability and the ability to operate under adverse conditions. A significant improvement in lifetime was obtained by the use of the oxide-coated cathode (due to A. Wehnelt) and developed by H. Arnold at AT & T By the middle of 1913, a filament life of 1000 hours had been achieved. A further advance, which took place somewhat later (although not successfully developed until 1930), was the indirectly heated cathode, which allowed the cathode to be heated from an alternating current supply, eliminating the need for a separate low-voltage DC battery supply. An important wartime development was the institution of mass-production techniques of manufacture. In the USA such firms as AT & T and GE and in Germany AEG produced valves on a very large scale. W.A. Atherton mentions that over one million were manufactured during the First World War.[11] Within six months of the United States's entry into the war, standardised valves were in production in that country, these being of the high vacuum type. In addition to valve production and technical improvements in the thermionic field, pure research was carried out in order to obtain a clear theoretical understanding of the device, and by the end of the war, a satisfactory understanding of the theory of vacuum tubes had emerged.

Following the war, domestic broadcasting expanded rapidly and this market soon began to absorb large quantities of thermionic triodes. By the end of 1922, there were thirty licensed broadcasting stations in the United States, and by 1924 there were five hundred. In Britain, events took place somewhat differently. In November 1922 the British Broadcasting Company came into existence. From this time onwards ownership of domestic radio receivers increased rapidly, both in the United States and Europe.

In both countries and on the Continent, the manufacture of vacuum tubes (principally for broadcasting) was well established by the early 1920s. The successful development of the vacuum diode and triode by this time resulted in the almost complete replacement of the crystal diode in radio receivers by the end of the decade. The era of the thermionic valve had now certainly begun.

2.2 Development of the tetrode and pentode

A major limitation of the triode was its inter-electrode capacitances, mainly between anode and grid, resulting in feedback and instability. As operational frequencies increased, this problem became progressively worse. Appleton states that A.W. Hull first suggested a screen grid to get rid of this effect,[12] although W. Schottky claims to have adopted this solution to the problem during the period 1916–19, whilst working in Germany. However, H.J. Round is generally credited with the effective development of the screen grid valve in 1926. The screen grid, placed between the anode and the control grid, had the effect of reducing the inter-electrode capacitance to a low value in the order of 0·001 to 0·01 pF. This device became known as the screen-grid tetrode. However at a certain anode potential this device suffered from the defect that, owing to the positive potential on the screen grid, electrons accelerated with sufficient velocity to cause 'secondary emission' of electrons from the anode, which were in turn collected by the screen grid. This effect resulted in the total anode current being reduced in value, and the screen current increased. If however the anode voltage was increased further, the anode received not only the primary electrons emitted form the cathode, but a further contribution from the screen. Under these conditions variation in anode voltage produced a nearly constant anode current. This was an advantage over the triode, because under conditions when anode current tends to be independent of anode voltage the possibility of amplification becomes very high, since the anode potential has almost no effect on the anode current, whereas the grid control potential is very effective. A further advantage of the tetrode over the triode was therefore the ability to achieve higher values of amplifcation.

The problem of secondary emission was solved by the invention of the pentode valve in 1928 by Telligen and Holst of the Philips Company in Holland.[13] This was done by the addition of a third grid held at anode potential, placed between the screen grid and anode. The effect of this grid—called the suppressor grid—was to capture the electrons released form the anode by secondary emission and thereby preventing the potential of the screen grid from rising due to this effect. The pentode valve had the advantage that, compared with the tetrode, it produced an almost constant current over a wider range of anode voltage, completely eliminating the 'kink' in the tetrode characteristic. Owing to its linear operating characteristics, the pentode became extremely popular during the period immediately preceding the Second World War. An alternative to the pentode, and possessing similar characteristics, was the beam tetrode. This device might be considered to be a tetrode in which the design has been modified so that the action of the suppressor grid is obtained by the insertion ofibeam-forming plates between screen grid and anode. By careful design, it is possible to achieve an even more linear operating characteristic with this device than with the pentode.

During this period before the Second World War, further developments in valve structure and geometry took place: double diode–triodes, double diode–pentodes and other multi–grid valves were manufactured to perform special tasks. In particular, triode hexode and triode pentode valves found an application as mixing valves in super hot receivers. However, these designs did not involve any new fundamental principle.

2.3 Impact of radar upon valve development

Owing to the advances described, the situation immediately before the outbreak of hostilities in 1939 was that the success of the thermionic valve was such that it appeared to hold a dominant and unchallenged position within the field of electronics as a signal amplifier and detector. The point-contact diode, now little more than an obsolete curiosity, like the coherer, had long been relegated to obscurity. However, with the advent of high-frequency radar, an unexpected challenge was to emerge.

A problem which arose during the early years of experimental radar was that, in order to build an efficient system, it was necessary to generate high powers and produce narrow beams. To achieve improved resolution of the target area, high frequencies were necessary. Specifically, what was required to achieve this was a device capable of operating at hundreds of megahertz with the ability to generate power of the order of kilowatts. The existing triode valve, in spite of development, was inefficient at these frequencies, a fundamental problem being that high grid power was needed to overcome electron inertia, and a consequence of this was that the valve, working at its design limit (or even beyond it) enjoyed an extremely short life. Interest during the early days of radar centred upon two possible alternatives, the klystron and the magnetron.

The triodes used in early experimental radar as generators of microwave frequencies depended upon the production of Barkhausen–Kurz oscillations. These oscillations were first noticed in 1919 by the aforementioned individuals when carrying out tests for the presence of gas in transmitting valves.[14] In this test, the grid of a triode was held at a high positive potential and the anode at a negative potential and an external circuit connected between grid and anode. The frequency of the resulting oscillations, usually greater than 300 MHz, depended largely upon the applied potentials. The lifetime of the valve under these conditions was short—usually a matter of minutes, owing to the grid melting! However, G. Marconi later developed (about 1933) a reliable valve using this principle, which had a life of up to 40 hours and delivered an output of about 5 watts at 600 MHz. The output was then amplified to obtain a final output of about 4 kW. Using this device Marconi himself very soon succeeded in demonstrating communication over distances of the order of 35 km on 600 MHz so convincingly that the Vatican authorities requested him to provide similar equipment for communication between the Vatican and the Summer Residence of his Holiness the Pope at Castel Gondolfo. This installation, the first microwave telephone in the world, was put into regular service in February 1933.[15] Nevertheless, by the outbreak of war, radar had not been installed in the Italian Navy and its absence could not have failed to have been an important factor in the action at Cape Matapan in 1941.

Meanwhile, work was proceeding in other countries. A.G. Clavier of the Laboratorie Central de Telecommunication in Paris started experimental work which was eventually to lead to the first commercial microwave link being established in 1933 between Lympne in England and St Inglevert in France.[16] In May 1940, Dr. M. Ponte of SFR Laboratories (France) brought to GEC Wembley samples of a resonant segment magnetron for pulse operation at 16 cm which gave pulse outputs of more than 500 W.[17]

In the United States, as early as 1921, A.W. Hull of the General Electric Company invented a form of magnetron, although it did not achieve any success at that time. Later, in the mid 1930s Bell Laboratories and MIT published theoretical work on microwaves. Page notes:

> 'an accidental observation by Taylor and Young in September 1922 that a ship interrupted some experimental high-frequency radio communication across the Potomac when it intercepted the propagation path between transmitter and receiver. The application was obvious to them (although unrelated to the experiment) and they immediately proposed that high-frequency radio-transmitters and receivers be installed on destroyers to detect the passage of other ships between any two destroyers in radio contact.'[18]

The first seagoing radar tests made in April 1937 were successful, and upon entry into the War the United States Navy possessed radar installations on 19 ships, apart from shore stations. Pierce in his paper 'The history of microwave tubes' states that 'Perhaps the first great step in understanding phenomena in microwave came with the invention of the klystron.'[19] Work was proceeding in this area in the late 1930s and papers were published by several authors including Brucke and Recknagel in Germany on the subject of focusing electrons in rapidly varying fields and by the Varian brothers in the United States, who described (in 1939) a klystron which they had invented.

Work in Japan was proceeding at Tohoku University under Professor K. Okabe and H. Yagi in the late 1920s and as early as 1931 S. Uda described in a published paper an experiment in which successful communication was established over a distance of 30 km on frequencies of the order of 600 MHz.[20] Prior to the attack on Pearl Harbour the Imperial Japanese Navy possessed radar using magnetrons as power generators.

In Germany, research in short-wave communication was also taking place at this time. For example, W. Pistor published a paper dealing with the reception of ultra short waves in 1930[21] and in 1935 the travelling wave microwave oscillator (a type of early magnetron) had been invented by O. Heil and A. Heil. By 1936, Germans working on the detection of aircraft possessed a device which could locate planes at a range of 80 km, and at the end of 1937, the first ship-borne radar was fitted on the pocket battleship *Graf Spee*. By 1939, Germany possessed a large number of radar sets. G.W.A. Dummer states that:

> 'though the German work began at least as early as in other countries it lagged behind after the outbreak of the Second World War. German policy was based on the assumption that the war would be short and consequently less effort was put into such basic work as radar than in Britain or the United States.'[22]

It was perhaps because of this policy that the Germans used triode transmission tubes during the Second World War, and did not develop the magnetron.

In Britain, the possibility of using radar waves to detect aircraft arose in the early 1930s and work on magnetrons started at the GEC Hirst Research Centre (Wembley) in 1931. Before the end of 1938, pulse outputs of the order of 1 kW had been obtained from glass magnetrons at about 30 cm.[23] Sir R.A. Watson-Watt, Superintendent of the Radio Division for the National Physical Laboratories, played a leading role in developing practical radar equipment and largely due to his work, the first radar station for the detection of aircraft was installed on the Suffolk coast in 1935. Following the success of this work, a chain of radar stations was installed. A highly important contribution was the invention of the cavity magnetron in 1939 by Randall and Boot at the University of Birmingham.[24] This device was superior to the existing klystrons in that its power output was considerably higher at centimetre wavelengths and represented a significant advance in microwave technology. Atherton states that Bell Laboratories had obtained pulsed powers of 2 kW at about 700 MHz using vacuum triodes at this time, by overheating the cathodes and operating the anodes at ten times their rated value, whilst the newly-invented cavity magnetron even at this early stage could produce 10 kW of pulse power at 3 GHz.[25] Pierce, in his History of microwave tubes, states:

> 'Although klystrons produced pulsed powers as great as 50 W by 1939, and EMI klystrons produced pulsed powers of around 30 kW during World War Two, as power sources they were overshadowed by magnetrons during that war.'[26]

It is interesting that a similar structure of a multicavity type was described by Alekseev and Malainov of the USSR during the same year.[27]

The first Soviet radar research station was set up by Oshchaphov at the Leningrad Electrophysics Institute and in January 1934 the first conference was held on locating objects by means of radar. Soviet physicists W. Muchin and O.I. Maljaniev constructed a magnetron, and an experimental radar was built that year. 'RUS 1' radar, constructed in 1937–8, was first used in the Russo-Finnish war (1939–40).[28]

The effect of the invention of the cavity magnetron was to limit the klystron to low-power applications, whilst the magnetron largely functioned as a generator of high-power oscillations. A further invention, somewhat later during the war, was the travelling-wave tube.[29] This device exhibited the characteristics of low noise and wide bandwidth and tended to be used in preference to the klystron for amplifying small signals, the latter being rather noisy and therefore unsuitable for this purpose. Klystrons were widely used during this period for laboratory oscillators and local oscillations in superheterodyne receivers.

2.4 Need for miniaturisation

The foregoing digression illustrates that work in the microwave field was progressing under the urgency of wartime conditions in all or most of the major belligerents. As operational frequencies increased, the efficiency of the thermionic valve as a detector of these very high frequencies became progressively poorer, adding to the urgency with which an efficient replacement needed to be found. The field of radar thus exposed the inadequacies of the conventional thermionic vacuum tube both as a receiver and generator of microwaves and stimulated work in two directions: firstly, towards new concepts of generating power at microwave frequencies, and secondly towards efficiently detecting these microwave signals. This need for efficient reception at microwave frequencies led to the development and large-scale production of miniaturised valves, since with their reduced inter-electrode capacitances they performed better at these frequencies. In addition, these valves had the secondary advantage of reducing the physical size and weight of electronic equipment, which was now becoming increasingly complex.

The first miniaturised valves appeared towards the end of the First World War as a result of the need for improved high-frequency performance, the earliest being the Marconi types V 24 and Q, which were designed to have very low inter-electrode capacitances. These valves were largely handmade and therefore expensive, and consequently found little commercial application.

A few miniaturised valves were manufactured in the inter-war years, but little development took place until the mid-1930s when Metro–Vickers produced the special types H 11 and L 11 intended for hearing-aid applications. From this time onwards, manufacturers in both Britain and the United States produced what were called 'midget' valves for this particular application. Early in 1940 the Radio Corporation of America marketed small thermionic valves specifically designed for portable radios. In Germany, miniature valves had been developed for military purposes well before 1939. With the advent of the Second World War, production of miniature valves for use at VHF and UHF frequencies rapidly expanded, particularly in the United States, and to a somewhat lesser extent in Britain. However, large quantities of standard-sized thermionic tubes were still being manufactured in both countries at the end of hostilities.

2.5 Conclusion

Compared with the situation immediately preceding the Second World War, the overall picture by 1945 was that of much greater diversification. Highly specialised, costly devices capable of generating extremely high powers existed at one end of the spectrum, and mass-produced low-cost miniature receiving valves at the other. It was in the latter area that semiconductor diodes had established themselves at the expense of the thermionic valve, as a detector of microwave frequencies. Later developments, involving the transistor, have since almost entirely replaced the vacuum tube as a low-power, low cost mass-produced device. It has not however been possible to replace the specialised high power high frequency transmitting devices, such as the klystron, magnetron and travelling-wave tube by any solid state equivalent.

2.6 References

1 STARLING, S.G. and WOODALL, A.J.: 'Physics' (Longmans, 2nd ed.) Chapter 53 p. 1198
2 'Encyclopaedia Britannica' 15th Edition, Vol. 6, p. 310
3 FLEMING, J.A.: British Patent No. 24850, 1904
4 Patent No. DRP 157 845. granted 13th January 1905. STOKES, J.W. '70 Years of radio tubes and valves' (Vestal Press)
5 DE FOREST, L.: 'The Audion: a new receiver for wireless telegraphy' *Trans. Amer. IEE*, 1906, **25**, p. 735
6 ATHERTON, A.: 'From compass to computer' (Macmillan, 1984), p. 197
7 *Ibid.*, p. 198
8 JEWKES, J., SAWERS, D. and STILLERMAN, R.: 'The sources of invention' (Macmillan 2nd Edn., 1969) p. 287
9 DUMMER, G.W.A.: 'Electronic Inventions 1745–1976' (Pergamon Press, 1977) p. 79
10 APPLETON, Sir E.: 'Thermionic devices from the development of the triode up to 1939' Lecture delivered before the IEE 16th November 1954 *in* 'Thermionic valves 1904–1954' (IEE, 1955) pp. 17–25
11 ATHERTON, W.A.: *op.cit.*, p. 199
12 APPLETON, Sir E.: *op.cit.*
13 DUMMER, G.W.A.: *op.cit.*, p. 81
14 PAGE, R.M.: 'The Early History of Radar' *Proceedings IRE*, **50**, pp. 1232–1236
15 ISTED, G.A.: 'Guclaimo Marconi and communication beyond the horizon, a Short Historical Note', *Proceedings IEE*, Jan. 1958, pp. 79–83
16 VOGELMAN, J.H.: 'Microwave Communication', *Proceedings IRE*, May 1962, p. 907
17 ALGER, J. and CLAYTON, Sir R.: 'Electronics in the GEC Hirst Research Centre. The first 60 years, *Radio & Electron Engineer*, **54**, 1984, p. 305
18 PAGE, R.M.: 'The early history of radar', *Proceedings IRE*, **50** pp. 1232–1236
19 PIERCE, J.R.: 'History of microwave tube art' *Proceedings IRE*, May 1962, p. 979
20 UDA, S., SEKI, T., HATAKEYAMA, K., and SATO, J.: 'Duplex radio telephony with ultra-short waves' *J. Inst. Electron Eng. Japan*, **51**, 1931, p. 449
21 PISTOR, W.: 'Reception of ultra-short waves' *Z. Hochfrequenztechnik*, 1930, **25**, p. 135
22 DUMMER, G.W.A.: *op.cit.*, pp. 73–74
23 ALGER, J., and CLAYTON, Sir R.: *op.cit.*, p. 305
24 BOOT, A.A., and RANDALL, J.T.: 'The cavity magnetron' *J. IEE*, 1946, **93**Pt, IIIA, pp. 928–938
25 ATHERTON, W.A.: *op.cit.*, p. 213
26 PIERCE, J.R.: 'History of microwave tube art' *Proceedings IRE*, May 1962, p. 979
27 ALEKSEEV, N.T. and MALAINOV, D.C.: 'Generation of high power oscillations with a magnetron in the centimetre band' *J. Tech. Phys. (USSR)*, 1940, **10**, pp. 1297-1300
28 ANTEBI, E.: 'The electronic epoch' (Van Nostrand, 1982), p. 171
29 KOMPFNER, R.: 'The travelling-wave tube as an amplifier at microwaves' *Proceedings IRE*, 1947, **95**, p. 124

Chapter 3

Historical survey of early research in semiconductors

Before discussing the growth of the semiconductor industry, which rapidly increased in scale following the invention of a solid state amplifying device, it will be useful to review the earliest technical developments leading up to that invention and without which such an event would have been, at least, highly improbable. Apart from an appreciation of the time scale involved, this survey may give some indication of how the accumulation of technical knowledge in this field progressed.

A further intention of this approach is to indicate the major factors affecting that growth and, in addition, outline how an industry manufacturing semiconductor devices of a non-amplifying nature came to be established.

3.1 The initial phase

Research in the semiconductor field began in the mid-nineteenth century and at first proceeded extremely slowly. This early work consisted of recording descriptions of physical phenomena involving the properties of what are now called semiconductor materials (i.e. materials with resistivities midway between insulators and metals). These first discoveries were carried out by individuals engaged in pure research, who described their findings by publishing papers in the existing scientific journals. It appears that at this stage theoretical explanations for the observations made were not attempted nor was any effort made to put these discoveries to practical use. A review of progress, listing discoveries in chronological order, now follows.

The first recorded effect that can be ascribed to semiconductor behaviour was the observation by M. Faraday in 1833 that silver sulphide, unlike metals, exhibited a negative temperature coefficient.[1] Soon after this, in 1835, M.A. Rosenschold, working in Germany, observed the phenomena of asymmetric conduction in solids[2] (this work was, however, neglected until its rediscovery by F. Braun in 1874).[3] In 1839 E. Becquerel described the effect of a voltage being generated at a junction between a semiconductor and an electrolyte when illuminated[4] (this was the first recorded instance of the photovoltaic effect). The next important recorded advance took place in 1850 when M. Faraday wrote a paper describing a temperature sensitive non-linear resistor.[5] (Devices of this type were manufactured at a much later date under the name of thermistors, being mainly used in the field of thermometry, and compensating for the effects of temperature upon resistance in electrical circuitry).

The photoconductive effect was first described by Willoughby-Smith in 1873, who noted that the resistance of a circuit element made of crystalline selenium decreased upon exposure to light.[6] This observation was later to lead to the construction of photoconductive cells (although at this time much work remained to be done before their efficiency was of a sufficient order to result in their widespread use).

In 1877 Adams and Day discovered the photovoltaic effect existing at a contact between selenium and a metal[7], that is, the ability of a rectifying contact to generate an electromotive force when illuminated, and in 1879 E. H. Hall described the effect bearing his name, i.e. that a voltage could be established between opposite faces of a crystal when placed in a magnetic field.[8] This effect was later to become of particular importance in determining the mobility of electrical charge carriers and whether conduction occurs by holes or electrons. It is by this means possible to decide whether the semiconductor under measurement is basically P or N type.

From the above it can be seen that over forty years had elapsed since Faraday's original discovery before the investigation of what have been described by Pearson and Brattain as the four main distinguishing characteristics of semiconductors i.e. the photoconductive effect, the photovoltaic effect, decrease in resistivity with increase in temperature and rectification.[9] It was only now that practical applications of the above work began to follow. During the period 1883–1886, C. E. Fitts, working in the United States, constructed a practical photovoltaic cell, by spreading a semi-transparent sheet of gold leaf onto the surface of a selenium layer on a copper backing plate.[10] Light penetrating the gold leaf to the contact between the gold and the selenium produced an electromotive force measured between the gold electrode and the backing plate. However, this early work came to nothing, and the development of the selenium photocell did not begin again until 1931. In addition to the above, he also constructed selenium rectifiers at this time, and in 1884 announced the discovery of an infra-red photovoltaic effect in naturally occurring lead sulphide (galena).

By 1904 J.C. Bose had made use of the non-linear rectifying properties of semiconductors to detect electromagnetic waves.[11] This highly important step was a significant advance upon previous methods of detection (i.e. the coherer, invented previously by Branley and developed by Lodge), and arose as a response to a definite need for an efficient non-mechanical device to perform this function.

The invention of wireless telegraphy acted as a stimulus, and led to an extensive exploration of many natural minerals as detectors of electromagnetic waves. During the first decade of the present century, many workers were engaged in this field, e.g. H. Dunwoody, who produced a practical detector[12], and G.W. Pierce, who between 1907 and 1909 constructed crystal rectifiers and also demonstrated that rectification effects were electrical rather than thermal, an important theoretical advance.[13] It was at about this time that G.W. Pickard first recognised that the element silicon could be used as a radio detector.[14]

These rectifying devices, constructed from naturally occurring minerals, became widely used before the advent of the thermionic valve as detectors of radio signals, and were popularly known as 'cats' whiskers'. The method of construction was simple, a wire being held in contact with the semiconducting material (for example, galena) by mechanical pressure, and this could be adjusted—as was often necessary—in order to obtain efficient detection, the device thus making use of the rectifying properties of the metal-to-semiconductor junction. (This device, now known as the point-contact diode, did not, however, receive any theoretical explanation of its action until the 1930s.)

Nevertheless, advances of a theoretical nature, which were to have important consequences later, were taking place. In 1900, M. Planck put forward a 'Quantum theory' according to which, radiation is not continuous, but only occurs in discrete units, or photons. In 1905, A. Einstein applied this concept to explain the photoelectric effect, the latter event stimulating investigation into the emission of electrons from solids.[15] Work in this field was particularly active at this time in Germany. By 1909, K. Baedeker had carried out the first work involving the systematic use of the Hall effect to study semiconductors[16] and J. Konigsberger published papers in 1907 and 1914 as a result of which the elements silicon, selenium and tellurium were classed as semiconductors.[17] [Germanium was added much later (1926). The first reference to that element being used as a detector was by E. Merritt, using what was, by later standards, highly impure germanium (95% purity).][18]

Although the point-contact crystal diode acted as an efficient detector, provided that occasional adjustment of the metal-to-semiconductor contact was made, it would not function as an amplifying device. A very real need existed at this time for a device that would not only detect but also amplify weak radio signals and operate without mechanical adjustment. This need was met successfully by the invention of the thermionic triode valve in 1906 by Lee de Forest[19], and largely due to its subsequent success as an amplifier, work in the solid-state did not make much progress, either theoretically or experimentally during the period from approximately 1910 to 1930.

Although the First World War stimulated developments in the field of thermionic valve technology, little work was done in connection with semiconductor materials. Also, because of the relatively advanced state of thermocouples and bolometers, most workers tried to use these devices for experiments being made in the field of infra-red detection. Nevertheless, work carried out using these types of detectors met with little success. However, one significant development was that of the Case 'thalofide' cell invented during the war. Case, working in the United States, discovered that if thallous sulphide was given appropriate heat treatment with the addition of oxygen, its sensitivity was considerably increased.[20] This cell had a fast response compared with that of thermopiles and bolometers and made possible the first military use of infra-red in guiding aeroplanes to landing fields and keeping convoys together.

Arnquist notes that this work was significant in being the first real attempt to enhance the natural photoconductivity of a material by the addition of small amounts of another substance.[21] Cells of a similar type were later made in Germany, Italy, Britain and the USSR.

Work during the First World War in Germany, carried out in the semiconductor field, appears to have been largely confined to the construction of optical systems for signalling purposes, operating in the visible spectrum and using selenium photoconductive devices.

3.2 Theoretical and experimental developments during the inter-war period

During the years following the First World War, indeed, during the whole inter-war period, interest in the possibility of constructing solid-state amplifying devices did not entirely disappear, in spite of the success of the thermionic valve, and efforts were made during this time by isolated individuals to construct successful amplifiers. These early workers were, however, usually hampered by inadequate resources and certainly by lack of theoretical understanding. For example, C. G. Smith working for Ratheon in the United States applied in 1923 for a patent in connection with a germanium current amplifier.[22] Important, although rarely quoted, work during the inter-war years was carried out by O.V. Losev in the USSR. Concerning this, Loebner writes:

> 'There can be little doubt that his production line of 'crystodyne' radio receivers, powered by 12 volt batteries, represents a thirty year headstart over the transistor radio. Why Losev's contributions were not included in prominent accounts of solid state amplifier history is an interesting question. After all, Losev's "Crystadyne" radios and detectors were exhibited during the mid-twenties at major European radio technology expositions and written up in German, French and Swedish trade publications.'[23]

However, in 1927, Losev turned his attention to the study of light emitting and light sensing diodes. In the same article as quoted above Loebner mentions the conversation with Professor B.A. Ostroumov, in which the latter states the reason for this was that difficulties arose regarding the purchase of zinc oxide crystals from the United States and, furthermore, vacuum tubes had become the dominant amplifying device by that time.

Various attempts were made during this period to construct a successful 'field effect' transistor. (This device is voltage controlled, unlike a binary junction transistor; i.e. by varying a voltage on a control grid insulated from a semiconductor, the flow of charge carriers in the semiconductor due to a voltage applied across it is modified by the charge induced by the voltage applied to the control grid. It is thus possible to control the flow of current through the device by this means. The efficiency of the device depends upon the degree of control of current obtained by varying the control grid voltage. Although the idea appears fundamentally simple, the actual construction of such a device in a useful form presents considerable practical difficulties.) Probably the best known of these attempts was that of J.E. Lilienfield, then Professor of Physics at Leipzig, who took out patents in 1926 and 1928 in which devices of this type were described, although it appears that they only offered operational concepts.[24] Whether these devices could have actually worked has recently been a matter for some discussion, although it appears to be doubtful.[25] In this context, Johnson states:

'Whether Lilienfield actually had an amplifier delivering useful output we do not know, and he certainly could not have used it in the radio circuit which he showed for illustration because of frequency limitations.'[26]

He questions V.E. Bottom's conclusion that the Lilienfield devices operated like injection transistors.[27]

Later, in 1935, O. Heil of the University of Berlin obtained a British patent for a proposed field effect transistor entitled 'Improvements in or relating to electrical amplifiers and other controlled arrangements and devices'.[28] This device consisted of a thin layer of semiconductor across which a potential difference was established by means of two electrodes at different potentials connected at either side. A third control electrode, insulated from the semiconductor, would then be used to modulate the resistance of the semiconductor layer, thereby controlling the flow of current in an external circuit connected between the two electrodes. The semiconductor materials proposed were somewhat unusual by subsequent standards, namely tellurium, iodine, cuprous oxide or vanadium pentoxide.

The first published record of experimental results in connection with a solid state amplifier was that carried out by Hilsch and Pohl in 1938.[29] In commenting upon this work, Pearson and Brattain write: 'The analogy with the vacuum tube diode also suggested to many workers of those days that what we really should do was to put a grid in the semiconductor diode and (Eureka!) the result would be an active triode with amplifying possibilities. The insurmountable block to this experiment was, of course, the technique of placing the grid in a region only 10^{-4} cm thick (the width of the space charge layer which could be calculated at that time). R. Hilsch and R.W. Pohl, working with alkali halide crystals in which the space charge layer could be made of the order of one centimetre in thickness, did put in a grid and made in principle an active device solid state triode circuit. The frequency cut-off of this experimental device was of the order of one cycle per second or less.'[30] (This low value of frequency cut-off was due to the extremely wide space charge layer). This experiment, although unsuccessful from the point of view of producing a 'usable' device nevertheless demonstrates that not only was it actually possible to build a solid state amplifier (although, because of its highly restricted bandwidth, useless for any practical purpose) but also that this work was guided by existing semiconductor theory.

Unsuccessful attempts to produce a solid state amplifying device continued during this period. In the United States, in 1938 or 1939, W. Schockley and A. Holden at Bell Laboratories tried to make a solid state amplifier using carbon contacts brought together under pressure exerted by a quartz crystal. Bardeen writes:

'They expected, but did not get, a usable output through change in resistance of the carbon as a signal was applied to the crystal.'[31]

Further unsuccessful experiments followed. Bardeen continues:

'Shockley's next step in December 1939 was to oxidise a metal wire screen thereby surrounding it with a semiconducting oxide, and attempting to limit conduction through the oxide from one side of the screen to the other. He thought that a negative voltage on the screen wires, like the grid voltage in a vacuum tube would control current flow through the oxidised screen. The idea was not successful.'

Describing the above experiment, the writer concluded:

'The theory on which these developments are based dates from the late twenties and thirties.'
'Nevertheless during this period agreement between theory and experiment was disappointing and at best qualitative. This is because semiconductors are what used to be called structure sensitive, that is, their properties depend critically on the presence of minute amounts of impurities or other crystal imperfections.'[32]

This fact certainly presented a major obstacle to progress and furthermore its importance was not fully appreciated at that time, largely because the significance of the role of minority carriers had not then been realised. In this context, Moll writes:

'most of the theories that were proposed prior to Shockley's definitive article contained some features in common with modern theory. However, the theories generally omitted the important role of minority carriers and for this reason can be considered of value only to the extent that they helped workers in the field to reach present-day concepts.'[33]

He continues:

'There is one very notable exception to the preceding statement viz a theoretical paper by Davydov in 1938.'[34]

Moll points out that Davydov's papers (there were two) were generally disregarded at the time, even though they were published in English in 1938. Pearson and Brattain write that:

'with the emphasis on the success of the space charge theory of rectification these two theoretical papers attracted little attention from other investigators and this blind spot regarding the role of the minority carrier continued to persist.'[35]

Certainly the lack of material of the required quality together with the lack of theoretical understanding doomed to failure efforts made during the inter-war years to invent a solid-state amplifier.

Although attempts during this period to construct solid-state amplifiers were unsuccessful, and in spite of what were later seen in retrospect as critical flaws in semiconductor theory, nevertheless important advances were made both in the understanding of the physics of the solid state and also in the technology of the manufacture of diode rectifiers and photocells.

Slightly retracing our steps in order to review developments on the theoretical front, in 1928, A. Summerfield, Felix Bloch and others applied quantum theory concepts to the theory of metals and in 1930 B. Gudden (in Germany) recognised that conductivity in semiconductors was almost entirely due to minute impurities present, rather than being a property of the bulk material itself.[36] By 1931 A.H. Wilson in England applied quantum mechanics in the investigation of the structure of semiconductor materials.[37] (Previously Schottky in Germany had put forward a semiconductor theory, but did not realise that holes would act as charge carriers.) Work carried out during this time established the basis upon which later theoretical work explaining transistor action was constructed. Subsequent work explaining the detailed behaviour of semiconductor devices therefore rests upon the concepts of post-classical physics. A theoretical picture was certainly beginning to emerge which could explain the movement of charge carriers in p-n junction semiconductors, and at semiconductor-to-metal contacts, by the beginning of the Second World War. The first model of a p-n junction was due to B. Davydov of the U.S.S.R. in 1938 detailing, as previously mentioned, the important role of minority carriers, and a paper describing a model for the metal to semiconductor junction was written by W. Schottky in the following year.

Nevertheless, considerable practical problems existed at that time, for instance, a major problem was that the semiconducting materials in the 1930s were too impure to provide an opportunity to link theory with experiment. (A method of obtaining a high degree of material purification, the zone refining process, was first described by the Soviet scientist P. Kapitza, working with Rutherford at Manchester in 1928. The significance of this method of extracting impurities was not however realised until W. Pfann, working at Bell Laboratories, perfected the process at a much later date.

3.3 Commercial and industrial applications during the inter-war period

Meanwhile, during this period, work was proceeding in the field of device applications. The copper oxide rectifier was introduced by L.O. Grondahl and Geiger in 1927 and was soon used extensively in such applications as low power rectifiers in battery chargers and wireless sets. Following this, in 1933, Grondahl and Geiger discovered the photovoltaic effect of the copper/copper–oxide barrier.[38] The selenium rectifier (invented, as mentioned earlier by C.E. Fitts) was developed shortly after the copper oxide rectifier, this work being re-opened by L. Bergmann and this cell proved in some respects to have superior characteristics, and as a consequence fairly rapidly replaced the copper oxide rectifier in many applications.[39] The main difference between the two types is that the copper oxide rectifier passes more forward current per contact area than the selenium rectifier, allowing a given rectifier current to be delivered by a smaller device than in the case of the selenium cell. However, the selenium rectifier can withstand larger reverse voltages (typically 20 to 30 volts per unit, whilst the copper oxide rectifier can withstand typically 6 to 8 volts per unit). Consequently the selenium rectifier is more efficient when rectifying current at high voltages, since fewer units need be connected in series in order to withstand a given voltage, thereby effecting a saving in space and weight. Both types however suffer from the disadvantage that they exhibit a fairly high resistance in the forward (conducting) direction, so that when large currents pass through them, excessive heat is dissipated, compared with that dissipated by the later germanium devices now currently used. This disadvantage renders them unsuitable for handling large powers. Large scale manufacture of selenium and copper oxide rectifiers began in the 1930s.

The re-introduction of the selenium photovoltaic cell by Lange, Bergmann and others in 1931 led to their use principally in connection with photographic exposure meters. Selenium photocells were also used on a limited scale during this period, mainly in the United States as automatic opening and closing mechanisms i.e. automatic doors in lifts and garages.

All the above applications found a market in situations where their competitor, the thermionic valve, was at a disadvantage for some particular reason. They possessed the advantage over the vacuum tube of not requiring a cathode heating element, were mechanically stronger and when connected in series, could stand large reverse voltages. Furthermore, photosensitive devices could be manufactured.

3.4 Military applications during the inter-war period

With political tension increasing throughout the 1930s both in Europe and the Far East, the possibility of military applications involving solid state electronics became correspondingly greater. Military interest appears to have been the main factor in stimulating renewed work, particularly in the field of infra-red detection. This work was not confined to one country.

The German infra-red military programme was resumed in about 1932.[40] Initially, the work was on lead sulphide, but soon came to include lead selenide and lead telluride. Infra-red detectors were made by both government research institutions and industry, supported by research in the universities. Experiments in increasing device sensitivity by cooling were carried out by using solid carbon dioxide as the coolant, and later, in a few cases, liquid nitrogen. [This work led to later wartime applications for these cells which included a ground-based device for detecting ships from the shore, a ground-based device for directing searchlights and anti-aircraft guns, a fixed base triangulation type rangefinder for night use, a Schnorkel installation for detecting ships and aeroplanes, airborne detection devices for aeroplanes and night targets, both air to air and air to ground, homing controls for missiles, and an infra-red proximity fuse. Towards the end of the war, lead sulphide and thallous sulphide detectors were used in connection with voice communication systems (known as 'Lichtsprechers').]

Work on lead sulphide was also being carried out in the years immediately preceding the Second World War in Poland at the University of Warsaw. Following the fall of Poland, members of the team engaged on this project found their way to Britain and continued to work in this field, being employed by the British Admiralty Research Laboratories at Teddington. This event initiated the British wartime effort in the area of solid-state infra-red device technology.[41]

Earlier work on infra-red detecting devices in connection with British air defence was carried out between 1935 and 1938 mainly by R.V. Jones. This work however involved the development of bolometers and thermopiles rather than thermoconductive devices and was terminated by the Air Ministry on 31st March 1938, priority then being given to work in the field of radar.[42]

In France, a pulsed radio system was fitted in the French liner *Normandie* (for the purpose of detecting icebergs) in 1936, and soon afterwards the military authorities began experiments in this area, although by 1939 no radar warning system had been installed.[43]

By 1939, Germany, Holland, Britain and the United States all possessed military radar apparatus, and consequently a need existed for crystal diode detectors of improved design in these countries.[44] Work was also underway in France, Italy and Japan.[45]

The introduction of radar at this time therefore stimulated renewed work in connection with crystal rectifiers and this became particularly important as the Second World War progressed, the reason being that thermionic valves did not perform well as frequency mixers and detectors of microwave frequencies. This problem became particularly important following the invention of the cavity magnetron, a highly efficient generator of high frequency oscillations, which allowed frequencies of 3 000 MHz (10 cm) to be used. (These higher frequencies were necessary in order to obtain better resolution of the target being viewed.) For the first time since the invention of the vacuum tube, an important need arose which could not be satisfied by that device. The fundamental limitation of the thermionic valve which gave rise to this situation was that the relatively large inter-electrode capacitances within the device introduced an operating time delay, and in addition the transit time of electrons between cathode and anode became significant at these high frequencies. These factors had the effect of limiting the upper usable frequency of the vacuum tube to about 4000 MHz, in spite of attempts to overcome the problem by reduction in the physical size of these devices. In contrast, the point contact crystal rectifier had an extremely low capacitance and the transit time of charge carriers was virtually insignificant. Consequently, point-contact transistors could operate at the required microwave frequencies successfully, and an urgent need arose for large quantities of these devices, it being important that these should be of constant quality.

In order to achieve satisfactory performance at microwave frequencies in radar applications it was however necessary to greatly improve the characteristics of point-contact diodes, and in view of the war situation, it was imperative to do this as rapidly as possible. Douglas and James point out that 'the important properties of a high frequency mixer are low conversion loss, low noise, high burn-out properties i.e. the ability to withstand current as experienced in modern radar equipment and uniformity of high frequency impedance'.[46] These requirements could only be met, given the existing situation, by a large-scale well-funded research programme, and this was embarked upon by Germany, Britain and, after 1941, on a large scale by the United States.

These requirements led to the abandonment of natural minerals and instead work was directed towards the controlled synthesis and purification of first silicon, and afterwards germanium. (In retrospect it appears that this work was to be vital at a later date, since without the ability to produce extremely pure semiconductor material, the construction of a semiconductor amplifier cannot be achieved by any presently known method.)

The first British solid-state detectors used in radar applications were manufactured by British Thomson Houston. These devices used commercial silicon of about 98% purity (extremely impure by subsequent standards) and showed considerable variation in sensitivity. In order to get consistent results further steps were taken to purify the silicon. This was first done by the General Electric Company, where silicon was purified to a higher degree than that commercially available at the time. Semiconductor theory predicted that the addition of minute amounts of certain impurities (i.e. beryllium and aluminum) would affect the rectifying properties of the crystal, and these were added, achieving some success, although the work proceeded largely on an *ad hoc* basis. (The effect of the addition of impurities at this stage was difficult to evaluate, since the added impurities themselves often contained further undetected impurities which were themselves significant.) In this context, Petritz states that 'even as late as 1948 rectifiers were made with tin-doped germanium which had a very high back voltage resistance for inverse voltages of 100 volts, and it was thought that tin was the doping element. We now know that impurities in tin, probably antimony or arsenic, were responsible for this action and that tin does not affect germanium electrically'.[47] The results from the British work assisted in stimulating further work in the United States, and it was this later work which was to be of crucial importance in manufacturing crystal rectifiers of high quality. Apart from producing crystal rectifiers of constant quality, the use of highly purified material led to greatly improved operating characteristics, both in the forward (conducting) direction (where substantial decrease in resistance was obtained upon the application of a voltage) and also in the reverse (blocking) direction, where reverse breakdown voltages were substantially increased and reversed currents minimised, resulting in the ability of the device to withstand greater voltage stresses, and also exhibit improved thermal stability.

3.5 Device developments during the Second World War

The entry of the United States into the war in December 1941 led to a great intensification of effort on the part of that country to produce large quantities of solid-state point-contact detectors of consistent quality, principally for use at microwave frequencies. Between about 1934 and 1940, work on semiconductor materials had been carried out under the auspices of Bell Telephone Laboratories, mainly in connection with silicon point-contact rectifiers, which were being manufactured on a limited scale. Apart from the previously mentioned unsuccessful attempts by Shockley and Brattain to produce a solid-state amplifier, a team of metallurgists and chemists, including R.S. Ohl, J.H. Schott and H.C. Theurer were carrying out goal-directed research on silicon with the object of improving short-wave detecting diodes. Owing to the urgent wartime need for radar detectors and mixers, the United States government instituted a large-scale programme both in the field of solid-state research and the manufacture of these devices. Co-ordination of effort was of great importance, J.H. Schaff and R.S. Ohl point out that 'the Office of Scientific Research and Development through the Radiation Laboratory at the Massachusetts Institute of Technology served as a co-ordinating agency for work conducted at various University, Government and Industrial laboratories in this country and as a liaison agency with British and Allied organisations.'[48] During the period 1941–2, the pattern for interchange of technical information was established, and work done both in Britain and the United States made available on a bilateral basis. Also, considerable product standardization was achieved during this period.

Although during the initial stages of work the British were somewhat ahead technically, the scale of American involvement was such that this situation did not long continue. E. Braun and S. Macdonald state that 'as many as thirty or forty US laboratories were delegated the task of examining semiconductors for use in radar.'[49]

Perhaps the most important research done in the field of germanium and silicon was that carried out at Purdue University and at Bell Laboratories. Work at the Physics Department of Purdue University began early in 1942 under the direction of K. Lark-Horovitz, a physicist who had previously worked in the field of nuclear physics. Efforts were concentrated upon the field of systematic study of germanium and its purification. Work was also carried out on device construction and germanium detectors were produced in the summer of that year. One important result of this work was Benzer's success in producing a high reverse back voltage rectifier. Discoveries regarding the physical properties of germanium were also made and the production of high purity polycrystalline germanium ingots achieved. The work done at Purdue during this time was later to be vital in providing the quality of semiconductor material which enabled transistor action to be feasible.

Meanwhile, Bell Laboratories largely concentrated their wartime effort on improving silicon point-contact diodes for use at microwave frequencies. Important fundamental work was also carried out, including that by J. Schaff and W. Pfann, who discovered the presence of Group III and Group V materials present in silicon and germanium. The first junction rectifying device had been discovered a little earlier by R.S. Ohl, about 1941. Pearson and Brattain write:

> 'One of the melts made by Schaff and Theurer was n-type at one end of the melt and p-type at the other, with rather a sharp boundary where the two meet. A section cut from this melt perpendicular to and including the interface boundary was found by Ohl to be an excellent rectifier and to exhibit a phenomenal photoelectromotive force. This was the p–n junction which has developed into a most important circuit element in our present-day semiconductor electronics.'[50]

The production of a silicon p–n junction was to have enormous consequences for the future solid-state amplifier, since without this technical advance, the junction transistor, upon which the industry was finally to depend, could not have been manufactured. Also, it now became theoretically possible to manufacture rectifying devices with much higher reverse breakdown voltages, capable of handling far greater power, and in addition multi-junction devices such as the thyristor, although these developments did not immediately follow. In this connection, J.M. Early writes at a somewhat later date (1962):

> 'Use of junction diode areas of a square cm or more, together with the provision for fixed cooling has resulted in units with peak current ratings measured in thousands of amperes. At the same time use of very thick junctions fabricated of high resistivity silicon has produced devices with peak voltage ratings of thousands of volts. Individual diodes rectify tens of kilowatts and rectifier combinations for conversion of a megawatt have been built.'[51]

During this wartime period, considerable government funding was made available for both research into the properties of semiconductor materials, and also the production of solid-state devices. A close connection existed between research institutions and industry; e.g. Petritz mentions that:

> 'F. Seitz of the University of Pennsylvania initiated a programme in connection with E.J. DuPont Nemours & Co. to develop a method to produce high purity silicon. K. Lark-Horovitz of Purdue University recommended Eagle Pitcher Co develop techniques to prepare high-purity germanium. J.H. Schaff and H.C. Theurer of Bell Laboratories investigated metallurgical methods for preparing ingots of germanium and silicon for use in point-contact studies.'[52]

Bell Laboratories constituted the research section of the American Telephone and Telegraph Company, of which Western Electric formed the manufacturing section. By 1945 Western Electric were producing twenty-six times as many rectifiers as in 1942, the production rate rising to over 50,000 devices monthly. Schaff and Ohl state that:

> 'This great increase was achieved simultaneously with marked improvements in sensitivity, the improvements in process techniques being reflected in manufacture by the ability to derive higher performance units in increasing numbers.'[53]

Although this survey must by its nature be incomplete, it is nevertheless possible to outline the general features of developments which took place in the semiconductor field from its earliest stages to the situation immediately prior to the invention of the point-contact transistor by Brattain and Bardeen in 1947.

The two main events influencing growth during this period were, firstly, the invention of the 'cats' whisker' crystal diode detector, which found an application in early radio, and secondly, the need which arose for an efficient detector of high frequencies in connection with the development of radar in the years immediately before the Second World War and which stimulated further development of the diode detector.

With regard to the original invention of the 'cats' whisker', the prospect of its widespread use in an important commercial—and possibly military—application stimulated work of an empirical nature in the field of semiconductor development. The inability of the crystal diode detector to act as an amplifier led to its neglect following the invention of the thermionic triode, which was immediately successful in this respect, and consequently both commercial and military development centred on improving the performance of thermionic valves.

During the inter-war years, little work was done, either theoretically or experimentally, except in two fields, which constituted a niche in which the thermionic valve could not compete, namely infra-red detection, where military applications were paramount, and also the development of the semiconductor diode rectifier, which had both commercial and military applications.

Wartime events completely transformed the semiconductor picture. Before the start of hostilities, government funded research was taking place on a relatively small scale, mainly in Britain and Germany, and largely confined to university and industrial research establishments. Some commercial involvement in material preparation took place and relatively small quantities of detectors were being manufactured, both for radar and for infra-red detection. By the end of the war, in only one country, the United States, a considerable industry had emerged, able to mass-produce large numbers of devices of consistently high quality. Important theoretical advances had been made and, specifically, the importance of purification of semiconductor material not only realised, but achieved in practice. Junction-type devices had been produced in addition to the point-contact type. A further achievement was that a number of strong research groups, including physicists, metallurgists and chemists, had been set up, together with a communication infra-structure involving Government, Industry and Research. Although no wartime attempts to produce a solid-state amplifier appear to have been made, the stage was now set for this highly important discovery.

3.6 References

1 FARADAY, M.: 'Experimental researches in electricity', *Bernard Quartier, (London)*, 1983, **1**, pp. 122–124
2 ROSENSCHOLD, M.A.: 'Experiments on the electrical Conduction of Solids,' *Ann. Pogg.*, 1985 **34** p. 437
3 BRAUN, F.: 'Resistance polarity in metal sulphides' *Ann. Pogg.*, 1874, **153**, p. 556
4 BEQUEREL, A.E.: 'On electric effects under the influence of solar radiation', *Comptes Rendus de l'Academia des Sciences*, 1839, **9**, pp. 711–714
5 DUMMER, G.W.A.: 'Fixed resistors (Pitman, London 2nd Edn. 1967) p. 147
6 SMITH, W.: 'The action of light on selenium', *J. Society Telegraph Engineers*, 1873, **2**, pp. 31–33
7 ADAMS, W.E. and DAY R.E.: 'The action of light on selenium' *Proc. Royal Society*, 1876, **25** pp. 113–117
8 HALL, E.H.: 'On a new action of the magnet on electrical currents', *Am. J. Mathematics*, 1879, **2**, pp. 287–291
9 PEARSON, G. and BRATTAIN, W.H.: 'History of semiconductor research' *Proceedings IRE*, 1955, **43** pp. 1794–1806
10 FITTS, C.E.: 'A new form of selenium cell' *Am. J. Science*, 1883, **26**, pp. 465–472
11 BOSE, J.C.: US Pat. No. 755840 (1904)
12 DUNWOODY, H.H.: US Pat. No. 837616 (1906)
13 PIERCE, G.W.: 'Crystal rectifiers for electric currents and electrical oscillations', *Physical Review*, 1907, **25**, pp. 31–60
14 PICKARD, G.W.: US Pat. No. 836531 (1906)
15 EINSTEIN, A.: 'A heuristic standpoint concerning the production and rransformation of light' *Ann. Physik*, 1905, **17**, pp. 132–148
16 BAEDEKER, K.: 'Using Cu I' *Phys. Zeitung*, 1909, **29** p. 506
17 KONIGSBERGER, J.: *Jb. Radio Act*, 1907, **4** p. 158 *Ibid.*, 1914, **11**, p. 84

18 MERRITT, E.: *Proceedings of NAT Academy of Science*, Washington, 1925, **11**, p. 743
19 DE FOREST, L.: 'The Audion: A new receiver for wireless telegraphy', *Trans. American IEE*, 1906, **25**, p. 735
20 CASE, T.W.: 'Thalofide cell—A new photo–electric substance', *Phys. Rev.*, 1920, **15**, pp. 289–292
21 ARNQUIST, W.N.: *Proceedings IRE*, 1959, **47**, p. 1420
22 SMITH, C.G.: US Pat. No. 1 679 448
23 LOEBNER, E.E.: 'Subhistories of the light emitting diode' *IEEE Trans.* 1976, **ED–23**, pp. 675–698
24 LILIENFIELD, J.: US Pat. No. 1745175. Filed 8 Oct. 1928 US Pat. No. 1877140. Filed 8 Dec. 1928 US Pat. No. 1900018. Filed 28 March 1928
25 WALLMARK, J.T.: 'The field effect transistor—An old device with new promise' *IEE Spectrum*, March 1964, p. 183. For further discussion, see JOHNSON, J.B.: Article in *Physics Today*, Feb. 1964, pp. 24–26.
26 JOHNSON, J.B.: Article in *Physics Today*, May 1964, pp. 60–62
27 BOTTOM, V.E.: 'Invention of the solid state amplifier', *Physics Today*, Feb. 1964, pp. 24–26
28 HEIL, O.: See Ref. 25. Also British Pat. No. 439 457
29 HILSCH, H.R. and POHL R.W.: 'Control of electron currents with a 3–electrode crystal and as a model of a blocking layer' *Zeits. f. Physics*, 1938, **III**, pp. 399–408
30 PEARSON, G.L. and BRATTAIN W.H.: *op.cit.* pp. 1794–1806
31 BARDEEN, J.: 'The improbable years', *Electronics*, 19 Feb. 1968, p. 78
32 *Ibid.*
33 MOLL, J.L.: *Proceedings IRE*, 1958, **46**, p. 1076
34 DAVYDOV, B.: 'On the photo electromotive force in semiconductors', *Tech. Phys. USSR*, 11938, **5**, pp. 79–86; and 'The rectifying action of semiconductors', *Tech. Phys. USSR*, 1938, **5**, pp. 87–95
35 PEARSON, G.L. and BRATTAIN, W.H.: *op.cit.* pp. 1794–1806
36 GUDDEN, B.: 'On electrical conduction in semiconductors', *Sitzungsberickle der Physmediz.*, 1930, **62**, pp. 289–302
37 WILSON, A.H.: 'Theory of electric semiconductors', *Proc. Royal Soc.*, 1931, **133**, p. 458; and **134**, p. 277
38 GRUNDAHL, L.O. and GEIGER, P.H.: 'A new electronic rectifier', *Trans. American IEE*, 1927, **46** pp. 357–66
39 BERGMANN, L.: 'On a new selenium barrier photocell', *Phys 2*, **32**, pp. 286–288
40 ARNQUIST, W.N.: *Proceedings IRE*, 1959, **47** pp. 1420–1439
41 Interview with Dr. D.H. Parkinson. (Late TRE Malvern), May 1985
42 JONES, R.V.: 'Infra–red detectors in British air defence', *Infra–red Physics*, 1961, **1**, pp. 153–162
43 DUMMER, G.W.A.: 'Electronic Inventions 1745–1956' (Pergamon P. 1977)
44 *Ibid.*
45 ATHERTON, W.A.: 'From compass to computers (San Francisco Press, 1984) p. 212
46 DOUGLAS, R.W. and JAMES E.G.: 'Crystal diodes' *Proceedings IEE*, 1951, **98**Pt.III, p. 160
47 PETRITZ, R.L.: 'Contributions of materials technology to semiconductor devices', *Proceedings IRE*, 1962, **50**, p. 1025
48 SCAFF, J.H. and OHL, R.S.: 'Development of silicon crystal rectifiers for microwave radar receivers' *Bell Syst. Tech J.*, 1947, **26** 1947, p. 1
49 BRAUN, E. and MACDONALD, S.: 'Revolution in miniature' (Cambridge University Press) p. 28
50 PEARSON, G.L. and BRATTAIN, W.H.: *op.cit.*, pp. 1794–1806
51 EARLY, J.M.: 'Semiconductor devices' *Proceedings IRE*, May 1962
52 PETRITZ, R.L.: *op.cit.*, p. 1025
53 SCAFF, J.H. and OHL, R.S.: 'Development of silicon crystal rectifiers for microwave radar receivers', *Bell Syst. Tech. J.*, 1947, **26**, p. 1

Chapter 4
Development of the transistor

This may be conveniently considered to have occurred in two phases. Firstly, the decade 1952–1962 encompassed the period from the invention of the grown junction transistor to the advent of the planar process and was one of rapid technical transition. During this time a wide variety of methods of construction followed each other in rapid succession. Furthermore, improvements in important electrical parameters such as reverse junction breakdown voltage, frequency response and power handling capacity established the transistor as a viable alternative to the thermionic valve, which, in spite of further miniaturisation, began to be replaced over an increasingly wide range of applications.

A fundamentally important development which occurred was the replacement of germanium by silicon as the basic semiconductor material for device manufacture. This change was made principally because of the ability of silicon to operate over a much wider temperature range, this factor being of particular interest to the military. The introduction of the planar transistor towards the end of the decade gave a further stimulus to the use of silicon. This was because the planar process could not be adapted to the fabrication of germanium devices.

The second phase began with the development of planar technology which established the basis for the development and manufacture of the integrated circuit, and already by the latter period of the decade was being used for the construction of both bipolar and field-effect transistors. It was at the end of this period, therefore, that the basic technology was developed upon which subsequent improvements in semiconductor manufacture have been made.

What follows is an account of the major advances in device technology, starting with the grown junction transistor and leading to the development of the silicon integrated circuit. Next, a brief survey outlines further developments which have arisen as a consequence of the impact of the integrated circuit.

The three final sections of this chapter consider the evolution of the small-signal semiconductor diode, the power rectifier and the thyristor.

4.1 Invention of the point-contact transistor

The first transistor was constructed on 23rd December, 1947 at Bell Laboratories by W.H. Brattain and J. Bardeen and announced to the press on 1st July, 1948, although patent application had been filed in February, 1948.[1] It was of a type which later became known as 'point contact' because of its form of construction.

This invention was the product of over two years of goal orientated research dating from the summer of 1945, when authorisation to work was given for a solid-state research programme, resulting in the establishment of the ultimately successful team, working under the direction of Dr. W. Shockley.

Two factors in the invention of the transistor have been mentioned by C. Weiner and are of importance. Firstly, he states:

'The significance of wartime semiconductor developments in setting the stage for the invention of the transistor cannot be over-emphasised.'[2]

and secondly:

'The semiconductor group was truly interdisciplinary, including experimental and theoretical physicists, a physical chemist and a circuit expert. There was also close collaboration with the metallurgical groups. This kind of interdisciplinary focus on a single basic research project would have been difficult to achieve in an academic environment at that time because of the traditional structure of the universities.'[3]

This interdisciplinary organisation was itself a direct consequence of wartime requirements and came about as a result of the demands of high quality semiconductor diode manufacture. Further progress in this field could only rest on a similar organisational structure. This type of organisation demanded considerable research and development backing and Bell Laboratories were perhaps unique in being able to supply this, employing at that time over two thousand highly qualified scientists, together with a considerable supporting staff.[4]

A third factor of importance at this stage was that the thermionic valve suffered severe limitations, particularly in certain fields; for example, in applications where requirements demanded minimum weight, size and power dissipation, together with high reliability. This problem was becoming particularly acute in the field of airborne equipment and rocketry, where components were subjected to severe mechanical stresses. There existed a need, therefore, especially in military applications, for a device capable of carrying out the function of a thermionic valve which was not subject to the limitations mentioned above. Military interest and involvement took place from the very beginning. Golding mentions that:

'Senior United States defence personnel received a preview of the transistor a week before the press announcement (which coincided with scientific publication) in July 1948.'

and following this:

'Funds were immediately allocated to expedite transistor development and production.'[5]

The invention of the point-contact transistor was the indirect result of a failure to produce a successful field effect device. Previous attempts to construct a solid state amplifier had been mainly concentrated on the latter type and Shockley had himself followed this path during the period immediately before the Second World War. His more recent attempts again proved unsuccessful. However, he writes, in discussing the sequence of events which led to the invention of the point-contact transistor, that:

> 'Our failure to make a transistor was creative. It led to research on the scientific aspects of Bardeen's surface states.'[6]

To explain their lack of success, Bardeen had suggested that electrons were being trapped at the surface of the semiconductor and consequently the application of an electric field at the surface would be ineffective in controlling the flow of charge carriers within the interior. It was, therefore, decided to investigate the surface properties of the semiconductor. This was done by probing the surface of the germanium crystal with two fine metal wires. During this work, Bardeen conceived the idea that it would be possible to construct a solid-state amplifier by carrier injection using one probe as a carrier injector (i.e. an emitter of charge carriers) and the other probe as a collector of charge carriers. To achieve this, theoretical calculations suggested that the point contacts had to be spaced extremely closely (about 2×10^{-4} in apart). Brattain writes:

> 'I accomplished it by getting my technical aide to cut me a polystyrene triangle which had a small, narrow, flat edge and I cemented a piece of gold foil on it. After I had got the gold on the triangle I could tell when I had separated the gold. That's all I did. I cut carefully with a razor until the circuit opened and put it on a spring and put it down on the same piece of germanium that had been anodised but standing around the room now for pretty near a week, probably. I found that, if I wiggled it just right so that I had contact with both ends of the gold, that I could make one contact an emitter and the other a collector, and that I had an amplifier with the order of magnitude of 100 amplification.'[7]

A solid state amplifier had therefore been created which worked on an entirely different principle to that of the field effect. Consequently, no difficulty was encountered by Brattain and Bardeen in obtaining a patent for the point contact transistor.

Although the apparatus used in the experimental work described above was relatively simple (the most complex piece of equipment being a cathode ray oscilloscope) the theoretical concepts behind this work were highly sophisticated, being based upon the application of quantum physics to the solid state. The semiconductor material chosen for this work was germanium, because its crystal structure had been investigated in some detail during the war years and also because of its relatively simple structure compared with compound semiconductors. Its lower melting point (960°C) compared with silicon (1430°C) simplified material preparation. Consequently, the first devices to be manufactured were also of the germanium point-contact type, although of somewhat improved construction, in which two metal electrodes or 'cats' whiskers' were mounted in contact with the surface of an N-type germanium single crystal wafer. In order to obtain transistor action, the sharply pointed contacts had to be mounted in close proximity to each other. By passing current, a 'forming action' took place, resulting in the area directly under each point contact being converted to P-type germanium, the assembly thus forming an N–P–N 'sandwich'. One contact was now forward biased with respect to the wafer, forming the 'emitter' and the other contact reverse biased with respect to the wafer, forming the 'collector'. The germanium crystal formed the third contact and was called the 'base'. Injection of majority carriers from the emitter probe initiated transistor action, resulting in a collector current, which, unlike that of the later junction transistor, could exceed the emitter current in value, producing a current gain, and thus transferring a current of equal or greater value from the low impedance emitter-base circuit to the high impedance collector–base circuit. These first devices provided power gains of about 20 dB (100) with 25 mW output at frequencies of up to about 10 MHz.

Initially, however, considerable technical difficulties existed. Individual devices differed widely in their characteristics and electrical instability leading to 'burn out' was not uncommon. In addition, noise levels were high, sensitive to changes in temperature and strongly affected by humidity. A further problem was that device action was extremely complicated, this being largely due to surface effects, making a theoretical analysis very difficult. Furthermore, this type of device construction did not lend itself readily to mass production methods, individual devices varying widely in their characteristics. As a result, production of point contact transistors began only on a small scale. Assisted by government contracts, production began at Western Electric in 1951, and by April 1952, about 8400 devices per month were being manufactured by that company. Also, by this time, several thermionic valve manufacturers, namely Ratheon, RCA, General Electric and Sylvania were manufacturing small quantities of point contact transistors, also under government contract.

4.2 Germanium grown junction transistor

Although the point-contact transistor was the first in the field, the junction transistor possessed far greater potential. This device was invented by W. Shockley working at Bell Laboratories and described in a paper published in July 1949, although Shockley names the date of conception of the device as early as 23rd June 1948 and wrote a detailed theoretical analysis in a patent application filed on 26th June 1948.[8] This device took relatively longer to develop successfully than the point-contact transistor. Shockley writes:

'In April 1949 some very inadequate "existence proof" amplifying devices were made.'[9]

These did not however demonstrate the technological potential of the junction transistor and he continues in the same article:

'a satisfactory realization of the junction transistor was not achieved until the Spring of 1950 and real excitement about it did not develop until early in 1951.'

The earliest grown junction transistors were constructed by selecting suitable specimens of polycrystalline material. However, late in 1948 G. Teal and J.B.

Little, also of Bell Laboratories, produced single crystals of germanium by a crystal-pulling process (known as the Czochralski process, and invented in 1918) in which P- and N-type impurities were successively added to the molten germanium in turn, allowing a single crystal to be drawn from the melt, in the form of a P–N–P or N–P–N structure. The crystal was then cut into a number of P–N–P or N–P–N bars, from each of which a junction transistor was fabricated, electrically conducting wires being attached to the three regions and the device then being encapsulated. However, this extremely important technical advance received little encouragement in its initial stages. In this context, Teal writes that this early work on crystal growing was done:

> 'without getting anyone's permission or approval and (we) acted only on our personal ideas.'[10]

and in the same article he states:

> 'Bill Shockley was opposed to the work on germanium single crystals when I suggested it, because, as he has publicly stated on several important occasions, he thought that transistor science could be elicited from small specimens of polycrystalline masses of material.'

Nevertheless, the successful growing of single crystal P–N junctions was to be the key to the success of the grown junction transistor, enabling it to be successfully fabricated. Teal writes in this context:

> 'Prepared the finest high-perfection single crystal P–N junction transistor patent no. 2 727 840 filed June 15th, 1950 granted to G.K. Teal December 20th, 1955.'[11]

A great advantage of the crystal growing process was that it enabled junction transistor devices with uniform characteristics to be manufactured in quantity. It is highly doubtful if this could have been realised by preparing junctions from polycrystalline material. This factor was to become of critical importance in the subsequent rapid development of the semiconductor industry. It would therefore appear that, although the junction transistor was undoubtedly the conception of Shockley, its practical realisation owed much to the work of Teal and Little.

Even in its earliest stages of development, the junction transistor was able to handle larger currents than the point-contact type and its noise level was considerably lower. J. O'Connor writes that:

> 'the junction transistor is 1000 times less noisy than the point-contact type.'[12]

Unfortunately its frequency response was somewhat inferior, owing to its relatively wide base width and opinion was by no means unanimous during the earliest years that it would entirely supersede the point-contact type, for this reason. For instance, Sparkes, writing in July 1952, states that:

> 'It should not be inferred that it will completely supersede the point-contact transistor.'[13]

However, the issue appears to have been definitely decided before the latter years of the decade. Shockley, writing at a later date, states that:

'In early 1951 it (the point-contact transistor) was displaced after the obvious superiority of the first microwatt junction transistor was clearly demonstrated.'[14]

Hibberd (writing in 1959) mentions that:

'the point-contact transistor suffered from inherent limitations in power-handling capacity and operating frequency; furthermore, it had a high noise figure and was liable to instability under certain conditions. As a result it had a relatively short life and can now be considered as obsolete.'[15]

Although, by 1952, Bell Laboratories were only producing about 100 junction type transistors per month, all of the germanium N–P–N type, already by this time a satisfactory theoretical analysis of its operation had been achieved, this situation contrasting with that of the earlier point-contact transistor. In this context, Sparkes writes:

'It placed transistor action on a quantitative basis. Measurement is the heart of science, and the junction transistor has brought the instrument to the stage when its performance can be described in numbers.'

Perhaps the greatest advantage of the junction transistor was that, unlike the point-contact type, reliable devices could be produced with constant, stable characteristics. In this context, Oberchain and Galloway write:

'Due to the difficulty of quantity production of transistors with uniform characteristics, immediate application to military equipments was impractical up to the spring of 1951. Evolution of the junction transistor in 1951 plus quality production ability with uniform characteristics for both these new-type low noise units as well as the somewhat older point-contact types, changed the somewhat restricted approach by military services to an active and expanding programme.'[16]

Certainly, the superiority of the junction transistor was rapidly appreciated by the military authorities, who decided to encourage its large-scale production at an early date. Bello mentions that Western Electric's million a month crash programme is designed solely to meet military needs. The military research and development had recognised in the fall of 1951, shortly after the announcement of the junction transistor, that the transistor was too vital to wait.[17] Military interest arose primarily because factors of space, weight and reliability were becoming of increasing importance, particularly to the US Air Force, as the complexity of electric equipment increased. Even before the invention of the transistor, the United States Armed Forces were actively concerned in the possibility of replacement of the thermionic valve by a smaller and more reliable device.

The advantages of the transistor were particularly important in the field of airborne equipment. Steutzer, in 1952, comparing the newly invented device with the thermionic valve, produced the data in Table 4.1:[18]

A consideration of the data in Table 4.1 shows that, apart from the decrease in size, weight and power requirements, all of which were highly desirable in

Table 4.1 Comparison of transistor with thermionic valve[18]

Stage Containing	Weight G	Size cm³	Power dissipation W	In flight h	On ground h
Standard tube	120	200	3	100	5000
Miniature tube	55	60	2·5	100	5000
Subminiature tube	35	40	2	500	5000
Transistor	15	10	0·2	75 000	75 000

view of the greatly increasing complexity of electronic equipment, the factor that perhaps stands out most is the vastly increased reliability of the transistor compared with the vacuum tube, particularly under normal operating conditions. In order to improve reliability, circuits using thermionic valves were forced to use systems involving parallel redundancy, this in turn increasing power requirements, weight and extra space required for the additional equipment. It is important to remember that these figures compare the characteristics of a newly invented device with that of one which had already undergone decades of development.

Drawbacks however remained. Transistors during their early stages of development could only handle low powers, and their frequency response was limited. It therefore became a matter of urgent necessity to improve their low value of frequency response, and also their power handling capacity. These limitations consequently stimulated the development of alternative methods of device fabrication.

4.3 Germanium alloy transistor

The main disadvantage of the early junction transistor when compared with the point contact device was its limited frequency response. Attempts were soon made to overcome this problem by alternative methods of construction, in order to achieve a narrow basewidth device. These efforts led to the invention of the Germanium Alloy junction transistor at the General Electric company in 1952. The production of P–N junctions by alloying was first described by Hall and Dunlap in 1950;[19] The first alloy junction transistors being described in the literature by R.R. Law, C.W. Mueller, J.I. Pankove and L.F. Armstrong[20] and in a separate paper by J.S. Saby.[21]

In this process, transistors of the P–N–P type were made by fusing indium emitter and collector electrodes into an N-type germanium crystal wafer acting as a base electrode. The emitter and collector electrodes consisted of indium (P-type) spheres which were mounted in physical contact with the germanium base. P–N junctions were formed by recrystallisation of the germanium from the alloy phase, this operation being carried out within an alloying furnace in an inert atmosphere at about 600°C. If the crystalline orientation of the germanium wafer was arranged so that the chosen crystalline axis was arranged to be parallel to its surface, alloying would take place in such a fashion that the junctions formed during the alloying process would themselves be parallel. During this alloying process the indium spheres melted and when subsequent recrystallisation took place during the cooling cycle, the recrystallised germanium contained a small amount of indium impurity, converting this portion of the N-type semiconductor into P-type. Following the alloying process, the base lead was attached to the germanium wafer by means of a metal disc coated with antimony-doped tin–lead solder. This sub-assembly was then spot-welded to the 'header', which consisted of a gold-plated metal (Kovar) substrate through which leads made electrical contact to the transistor, these leads being electrically insulated from each other and from the header by glass insulation. Tin-plated copper wires were now attached to the indium heads and spot welded to the ends of the leads projecting from the top of the 'header'. Encapsulation was completed by welding a small metal can onto the top of the header, totally enclosing the device. This process was completed in an inert atmosphere, usually argon or nitrogen. The can itself was often filled with a substance, such as 'bouncing putty' or alumina which had a marked effect upon the electrical characteristics of the device. This fact illustrated a basic weakness in this type of design, its tendency to electrical instability, due to the unpassivated nature of the surface of the junction, causing electrical parameters to vary during use. This device, however, exhibited a significant improvement in frequency response and current gain when compared with the grown junction transistor. This arose because the alloying process enabled much finer control of basewidth to be achieved through controlling the profile of the alloying furnace. This requirement for accurate control of furnace temperature led in turn to improvements in furnace construction, which were to be of great value not only in the development of alloy junction transistors, but also in the subsequent development of diffused transistors. A great advantage of the alloy junction transistor was that it was possible to adapt its fabrication to large-scale production, enabling it to be produced at less cost per unit than the grown-junction device.

Various modifications of the germanium alloy process were attempted in order to obtain increased frequency response; for instance the Philco company introduced, in the autumn of 1953, a jet etching technique in which the germanium wafer was etched by an electronically controlled jet of electrolyte. As the thickness of the wafer decreased, its transverse resistance correspondingly increased. By monitoring the process, the desired thickness would be obtained. Metal contacts were then plated to both sides of the thin section of the wafer. When this structure was heated, alloying into the narrow base took place. The basewidth of these devices was typically 0·0002 in compared with 0·001 in for a conventional germanium alloy junction transistor. This device however suffered from the disadvantage of mechanical fragility and was relatively expensive, since production yields tended to be low, owing to manufacturing difficulties. This type of device is usually described as a surface barrier transistor.

A fundamental difference between the alloy junction transistor and the grown junction device was that of the method of construction, the alloy junctions being of the abrupt or 'step' type, whilst the grown junction devices being of the 'graded' type. The emitter–base step junction gave the alloy device an advantage of high emitter frequency, improving the current gain of the device, particularly since the emitter itself possessed a low resistivity. The alloy junction device was more efficient as an electronic switch than the grown junction transistor and for this reason was widely used in computers towards the latter end of the decade. A disadvantage was, however, that the abrupt junction between the collector and base resulted in a higher capacitance per unit area when compared with the grown junction device, this tending to limit the high-frequency response. The effect of this collector–base capacitance was also to cause instability when the device was used as an amplifier, unless special precautions were taken to prevent this occurrence.

A.A. Shepherd gives typical values for alpha cut-off frequency as follows:[22]

Conventional alloy transistor	5–10 MHz	(P–N–P type)
Surface barrier alloy transistor	50 MHz	(P–N–P type)
Grown junction transistor	1–10 MHz	(N–P–N type)

H.L. Morton gives typical values for alpha cut-off frequency for a point-contact transistor as up to 50 MHz.[23]

Towards the end of the decade, a wide variety of devices were being manufactured on a fairly small scale with the aim of extending frequency response, although they all suffered from various disadvantages, for example, Shepherd writes in 1957 that germanium P–N–I–P and N–P–I–N (I = intrinsic) transistors were being made with alpha cut-off frequencies up to 200 MHz.[24] However, like the other alloy devices so far considered, power output was low, being limited to about 100 mW at these frequencies. Their method of construction was such, however, that large-scale production was not feasible and all these various attempts to produce an alloyed junction high frequency transistor soon gave way to diffusion techniques, which allowed much more accurate control of junction depths, whilst producing a graded junction structure. Nevertheless, grown junction and alloy junction devices continued to be produced on a mass-production basis well into the late 1960s.

4.4 Silicon grown junction transistor

The silicon grown junction transistor first entered production in May 1954, being made by Texas Instruments Inc. of Dallas, Texas. It was the first successful silicon transistor and the key to its success was the ability of that company to produce high quality monocrystalline silicon.

These devices were manufactured in a similar fashion to the earlier germanium grown junction transistors, using the Czochralski crystal growing technique. Using this method, the emerging crystal was pulled from the silicon melt, being successively doped with P- and N-type impurities, this resulting in a series of single crystal N–P–N sandwiches which could then be cut up into a number of N–P–N bars, assembled into gold plated Kovar headers and then encapsulated. Like the germanium grown junction device, the surface remained unpassivated and leakage currents were correspondingly high. Owing to the higher melting point of silicon (1430°C) compared with germanium (960°C), additional contamination problems arose during the crystal growing process. These difficulties were largely overcome by such means as the use of fused quartz crucibles rather than graphite in the crystal pulling operation. This device (described by G. Teal and by Adcock, Jones, Thornhill and Jackson) had the disadvantage that, like the germanium grown transistor, it was difficult to fabricate with small area junctions and a narrow basewidth; consequently its frequency response was limited to a few hundred kilohertz.[25,26]

The silicon grown junction transistor was largely the work of G. Teal, who joined Texas Instruments (TI) from Bell Laboratories in November 1952 and established the TI Central Research Laboratories. He announced his success at a meeting of the IRE National Conference of Airborne Electronics in Dayton, Ohio, on 10th May 1954. Referring to this meeting, Teal writes:

'During the morning sessions, the speakers had unwittingly set the stage for us. One after the other they had remarked how hopeless it was to expect the development of a silicon transistor in less than several years.'[27]

Teal then made his announcement, and moreover carried out an experiment, demonstrating to the audience the superior high temperature properties of the device. Teal continues:

'The silicon transistor was the turning point in TI's history, for with this advance it gained a big head start over the competition in silicon transistors until 1958. TI's sales rose almost vertically. The company was suddenly in the big league.'[28]

In this context Golding writes:

'For three years TI was effectively the sole supplier of silicon transistors at prices yielding profit margins which the chairman of the company later described as exceptional.'[29]

The silicon grown junction transistor was of particular interest to the military, owing to its ability to operate at higher temperatures than germanium devices which were unsuitable for operations involving rocketry. Indeed, large scale military involvement in semiconductors really dates from the invention of the silicon transistor. Sangster writes:

'when Sputnik went up, the US had the electronic miniaturization capabilities that allowed the nation to get back into the space race, even though our available rocket boosters were vastly less powerful than those of the USSR. We didn't have enough rockets to get the vacuum tubes into space—they were too heavy.'[30]

A pressing need therefore existed at the time, at least in the field of rocketry, for a viable alternative to the thermionic valve. Certainly no time was lost in getting the silicon transistor into production, as can be seen from the following dates, given by W. Adock:[31]

12th March	Memo predicting that transistor could be made from current silicon material.
13th April	N–P–N junction crystal grown
14th April	Working transistor assembled
8th May	Announcement that TI was in production on Si transistor
14th May	G. Teal's announcement at Dayton, Ohio.

This time scale may itself illustrate the strength of TI insofar as this company has always relied on a policy of continual development of the product whilst actually in volume production.

The success of the silicon grown junction transistor, with its wider operational temperature range, played an important part in pointing the industry in the direction of silicon, although germanium low power devices continued to be made in large numbers for a further decade.

4.5 Mesa transistor

Although diffusion techniques had previously been used in the manufacture of transistors, the discovery of the oxide masking process by Frosch and Derrick[32] led to the development of the Mesa transistor, described by Aschner, Brittman, Hare and Kleinach.[33]

This device was a significant advance on previous types, largely due to the precision with which junction depths could be controlled by diffusion. This allowed a far greater degree of control of basewidth resulting in improved high frequency performance. Sparkes mentions that, using the diffusion technique in conjunction with mesa devices, basewidths of less than 5×10^{-4} cm could be achieved, about ten times less than was previously possible.[34] This resulted in cut-off frequencies in excess of 100 MHz.

Typically, devices made by this method would be fabricated upon the surface of a single-crystal N-type slice. Firstly, a P-type impurity layer would be diffused over the whole of the slice surface. The N-type substrate would constitute the collector and the P-diffusion the base. A second diffusion of N-type impurity would now convert the surface of this layer back to N-type, an N–P–N diffused 'sandwich' thereby being formed. It was important that this second N-type diffusion penetrated much more rapidly than the first, in order to prevent further diffusion of the P-layer during this operation. The next stage was to use oxide masking and photoetching techniques to remove a portion of the top of the slice, producing the characteristic 'mesa'. This was necessary because otherwise the P–N junction areas were too large, resulting in high depletion layer capacitance and therefore degrading high frequency performance. It was now necessary to attach the base and emitter leads. Attachment of the base lead was particularly difficult with this type of construction and various methods were attempted including the alloy diffused method already described.

Note that the working area of the mesa transistor is confined to a thin layer at the surface. In order to maintain a high breakdown voltage between base and collector it was necessary that the collector resistance was relatively high. Because of the thickness of the collector region, series resistance was also high and this had the disadvantage of reducing the power-handling capacity of the device. In addition, this high series resistance formed a time constant with the collector–base depletion layer, adversely affecting its switching characteristics. Any attempt to reduce the thickness of the collector resulted in the extreme fragility of the slice, lowering production yields. This problem was solved by using the epitaxial process.

The advent of the planar process has not completely supplanted the mesa method of construction which has continued to be used whenever very high reverse breakdown voltages are needed, typical applications being large power transistors and silicon controlled rectifiers. Unlike the planar devices, where reverse junction breakdown voltages of only a few hundreds of volts can be achieved, the mesa approach enables breakdown voltages of several thousand volts to be reached, and this process has consequently remained the standard method of attaining very high reverse junction breakdown voltage devices.

4.6 Planar transistor

The invention of this process by J.A. Hoerni of the Fairchild Corporation in 1959, and first described in Europe by V.H. Grinich and J.A. Hoerni in February 1961, constituted a significant advance in semiconductor device technology, resulting in the fabrication of devices with improved reliability and electrical characteristics, together with decreased cost, and in addition opening up the prospect of successful integrated circuit manufacture.[35] With the exception of certain applications, such as high voltage rectifiers and thyristors, the planar technique rendered all previous methods of device construction obsolete. The process was described as planar because it resulted in the junctions being covered by an electrically stabilising oxide, producing a flat, or planar, surface.

The effect of this major technical breakthrough may be gathered by the following remarks made by J.J. Sparkes who was present at the European Conference in Paris in 1961 when the details were made public:

'Victor Grinich of the Fairchild Corporation presented graphs of the change with time of current gain, base–emitter voltage and cut-off current of planar transistors which were so much better than anyone had seen before that it was quite obvious that if they were genuine a real breakthrough had been achieved. After several hours' discussion with Grinich it became clear to me that the planar process was the process of the future. It was an unpalatable conclusion, since, just at that time, many companies had recently invested large sums of money in the double-diffused, the alloy-diffused or micro-diffused process with the hope of achieving a clear production run of a few years.'[36]

The planar device was developed directly out of the silicon mesa process. Dummer writes:

> 'At Fairchild, Dr. J.A. Hoerni, a physicist, was trying to develop a family of double-diffused silicon mesa transistors. But instead of mounting the base layer on top of the collector, the traditional mesa approach, Hoerni diffused it down into the collector and protected the base–collector junction on the top surface with a layer of boron and phosphorus diffused silicon dioxide. This first planar transistor was less brittle than the mesa and far more reliable–dust or other foreign matter could not contaminate the P–N junction.'[37]

The advantage of growing a layer of silicon dioxide over the junctions was that for the first time a satisfactory method of electrically stabilising the surface of the device had been discovered. Reverse leakage currents were greatly reduced by this means (typically from a tenth to a hundredth of their previous value); consequently thermal variations were proportionally reduced. Also, long term stability was significantly improved and reverse breakdown voltages increased. The net result of these changes was that, not only were characteristics upgraded, but devices became more reliable. The planar process resulted in batch production methods being greatly facilitated, and reproducibility of electrical characteristics was greatly improved. Owing to the nature of this process, devices nearest to each other on the slice tend to have similar characteristics, enabling 'matching' of electrical parameters to be carried out. Another advantage was that of increased flexibility, since 'tooling up' to produce a different planar device could be done at minimum cost and with minimum disruption, merely involving the changing of contact masks, diffusion profiles and possibly the resistivity of the starting material and doping levels. The planar process has not, however, been widely used to manufacture germanium devices, owing to the difficulty in growing a stable oxide onto a germanium surface and this fact further accelerated the trend towards the use of silicon rather than germanium as the basic material for semiconductor manufacture. Nor was it possible to produce high power devices which therefore continued to be manufactured principally by the mesa process.

In conjunction with the invention of the planar transistor, Hoerni and Noyce also developed the technique of depositing metal contacts upon the surface of silicon and this was to be of significant importance in the later development of the integrated circuit, allowing electrical interconnections to be made between devices on the same wafer. Contacts of this type are now universally used, aluminum being deposited upon the surface of the slice by the vapour deposition process, unwanted aluminum then being removed by photomasking and etching techniques. Contacts from the chip to the header are usually made by thermocompression bonding, this technique being developed at Bell Laboratories during the mid 1950s by O.L. Anderson, H. Cristensen and P. Andreasch.[38] This technique consists of bringing another metal, usually gold (melting point 1063°C), in contact with the aluminum bonding pad (melting point 660°C) in a reducing atmosphere under pressure. A gold–aluminum eutectic alloy is formed at a temperature of about 350°C and when cooled forms a strong reliable bond. Alternatively ultrasonic bonding using aluminum wire may be used.

Apart from the more recent introduction of ion implantation techniques the process so far described has, with minor modifications, remained essentially the standard method of transistor and integrated circuit manufacture to date, although great advances in micro-miniaturisation and automation have taken place.

4.7 Early development of the power and frequency characteristics of the transistor (1950–70)

During the period 1950 to 1970, the thermionic valve was replaced in an ever increasing variety of applications, firstly in the military field and then somewhat later in industrial and commercial applications. The rate at which this replacement occurred was due to both technical and financial factors. Although it is convenient to separate these factors for the purposes of analysis, it should be remembered that they are closely interrelated, the aim of the device manufacturer always being to achieve an improved product at decreased cost.

What follows is a brief survey of technical developments within the stated period, concentrating upon two key parameters, with the intention of demonstrating the rapid rate at which one improvement followed another in these particular areas.

In discussing the technical development of the transistor, perhaps the most illustrative and important factors to consider are those of frequency response (alpha cut-off frequency) and power handling capacity. Unfortunately a conflict exists between these two parameters. In order to fabricate a device with a high frequency response, a narrow basewidth is necessary. This is because base thickness imposes an upper limit on frequency response, owing to the finite transit time of charge carriers flowing across the base from emitter to collector. Another important requirement is that series base resistance and collector depletion layer capacitance should be low. This implies that the junction area should be as small as possible, thereby limiting the power handling capabilities of the device.

Because of these conflicting requirements it is convenient to consider the development of the power transistor separately from that of the high-frequency small-signal device. It should not be imagined however that, in the design of either type, considerations of the other parameter mentioned above should be neglected. It should also be noted that a wide spectrum of applications exist in which a combination of both high frequency response and power handling capacity are desired and consequently devices with a wide range of characteristics have been produced in order to meet these requirements.

4.8 Development of the power transistor

The first transistors were of the small-signal type, capable of handling only about 50 to 100 mW maximum. However, very soon attempts were made at producing devices capable of dissipating larger powers. Shepherd mentions that power transistors (usually defined as those capable of operation above 100 mA) first made their appearance about 1952, and were of germanium alloy construction.[39] They were similar to the existing low power type, but the collector was mounted on a copper heat sink, which could be screwed to a chassis, allowing about 10 W dissipation.

A typical early power device was that described by Roka, Buck and Reiland in August 1954.[40] This was a developmental germanium power transistor with a collector dissipation of 20 W at room temperature and 8 W at 80°C, delivering a peak collector current of about 1 A at a peak collector voltage of 60 V. The device was stud mounted on a copper base, insulated from the chassis mounting by a mica disc, the electrical lead terminals being taken out of the top of the header can.

Certainly the rapid advance in power handling capability of semiconductor devices during the first decade of transistor development testifies to the amount of research and development effort carried out in this field. Hibbert supplies the figures in Table 4.2:[41]

It must be noted, however, that these early germanium power transistors were all low frequency devices (typical common-base cut-off frequency 100 kHz) and were restricted to relatively low operating temperatures (approximately 70 to 80°C).

Improvements which followed in 1955–57 included the addition of gallium or aluminum to the indium emitters, which resulted in higher current gain. In the latter year the germanium diffused base power transistor was developed, which had a lower base resistance than the alloy type, resulting in increased power gain. (For example, 5 Ω compared with a previous value of about 30 Ω.) A further development in 1963 was the epitaxial base transistor. Because of the epitaxial design, base resistance was reduced to a value of about 1 Ω, resulting in both further current and power gain.

Technical advances during this period included interdigital geometry, this involving a larger emitter periphery for a given emitter area, thus improving emitter efficiency, and therefore current gain. Another important innovation was the 'overlay' structure introduced by RCA in 1965, the geometry in this case consisting of a number of separate emitters within a common base structure. Shepherd mentions that between 1965 and 1970 the development of

Table 4.2 Advances in power handling of semiconductor devices[41]

Year	Material	Device	Collector power dissipation
1950	Germanium	Grown junction	100 mW
1952	Germanium	Alloy junction	200 mW
1953	Germanium	Alloy junction	20 W
1954	Germanium	Alloy junction	100 W

this type of device increased output power capacity from 2 W to over 100 W at a frequency of 2 GHz.[42]

The silicon power transistor was introduced in 1957. An advantage of silicon over germanium is its ability to operate at higher temperatures (up to about 150°C). The first silicon power devices were N–P–N alloy types, with breakdown voltages of up to 300 V, able to dissipate up to 80 W, and with a cut-off frequency of about 1 MHz, this frequency being significantly higher than that of the available germanium types. Current gain was however somewhat below germanium power devices in value.

The planar approach does not appear to have been used during this period, although the triple-diffused mesa structure, dating from about 1965, made it possible to obtain reverse bias voltages in excess of 1000 V.

During the 1960s most power devices were encapsulated in hermetically sealed packages, such as TO 3, TO 36 or TO 66 headers. During the mid-1960s, however, plastic encapsulation began to be introduced in order to reduce header costs, which by then, owing to the improved production efficiency of devices, were becoming an appreciable proportion of the total cost of the device.

It can therefore be seen from the above remarks that during the period under discussion, extremely rapid advances in power handling capacity of semiconductor devices took place. This had the effect of displacing thermionic valves from such important market areas as power output stages in radio receivers and small transmitters, and towards the end of the period solid-state devices were being used in television receivers for such applications as raster deflection. Indeed, Overton describes a complete television receiver using transistors as early as 1958.[43] Other areas in which thermionic valves were being supplanted at this time included telemetry systems, mobile communication systems and radar applications. The increase in cut-off frequency of power transistors which took place in the latter half of the 1960s contributed greatly to the expansion of the use of solid-state devices into the field of communications.

4.9 Development of transistor HF response

The first transistors to be manufactured were of the point contact type. Morton quotes frequency response of these devices as being 5 MHz in September 1949, this figure being raised to between 20 to 50 MHz by January 1952.[44] Point contact devices, however, suffered from instability, high noise levels and extremely low power dissipation and were soon supplanted by grown junction and alloy junction devices. These devices could, however, only achieve much lower values of frequency response. Shepherd in 1957 gives typical values of cut-off frequency for germanium grown junction devices as between 1 and 10 MHz.[45] Their main frequency limitation was due to the high series resistance of the thin base layer. This problem was overcome by a modification of the basic process, the grown junction tetrode (dating from 1953), which involved applying a second connection to the base layer. By applying a few volts bias between the base leads, the current flowing from emitter to collector was constrained to flow in such a fashion that the effect of base series resistance was reduced. This resulted in a high-frequency cut-off of about 50 to 100 MHz, and suffered from the disadvantage of the need to provide an additional potential for the second base electrode.

High-frequency germanium alloy junction transistors were fabricated by reducing the size of the device, using collector and emitter dots of about 0·15 and 0·01 in respectively, and a base width of 0·0005 in in order to reduce the value of the depletion layer capacitances. Cut-off frequencies of up to 10 MHz were attained by using this approach.[46]

A further modification of the germanium alloy process was the introduction of the alloy drift transistor. This device had a non-uniform base resistivity, producing a field which swept the charge carriers from emitter to collector at a higher speed than in the case of the uniform-base type, thereby reducing the transit time and increasing the frequency response. This modification resulted in a maximum cut-off frequency of about 30 MHz.

A more successful attempt at increasing the frequency response of the germanium alloy transistor was the micro-alloy diffused transistor (MADT) described by Thornton and Angell, who describe the fabrication of high-speed switching transistors with cut-off frequencies of 125 MHz and 250 MHz, and a device capable of 1 W output at 70 MHz.[47] This form of construction, considered as a significant advance at the time, was, however, soon eclipsed by the invention of the diffused mesa and later planar transistor, this process allowing not only accurate control of basewidth, but which was, unlike the germanium alloy process, a batch method of production, ensuring a much greater uniformity of device parameters, together with a substantial reduction in device production costs. Also, compared with the micro-alloy diffused transistor, leakage currents were lower (particularly in the case of the planar process) and the device was considerably less mechanically fragile.

Hybrid versions of the alloy and diffusion process were attempted during the late 1950s and early 1960s, an example being the post-alloy diffused transistor, described by Beale. This device exhibited cut-off frequencies of up to 200 MHz.[48]

However, the diffusion process, used first with mesa and later planar devices, soon supplanted all previous types. In 1956, germanium diffused junction transistors were being manufactured with cut-off frequencies of 500 MHz and silicon units with cut-off frequencies of 100 MHz. By 1959, germanium mesa transistors were being produced with extremely thin base layers (for example, typically 0·00005 in) resulting in cut-off frequencies in excess of 1000 MHz.[49]

Although this rapid improvement in frequency response over barely one decade appears to be impressive, it should be noted that thermionic valves were being produced during this time with cut-off frequencies in excess of 10 000 MHz (10 GHz).[50] It was not until the latter 1960s that this figure began to be approached, using planar manufacturing techniques.[51]

4.10 Field effect transistor (FET)

The invention of the field effect transistor (FET) predates that of the bipolar device. The first recorded attempt at producing a multi-layer structure of this type was made by J. Lilienfield, then professor at Leipzig, who filed patents in

1926 and 1928. This early work, followed by that of Pohl and Heil, has been described elsewhere.[52] These efforts were significant not only because they determined the direction in which later developments were to proceed, but also because the field-effect principle had been understood and described, although practical realisation had yet to be attained. In this context Braun and Macdonald write:

> 'Brattain and Bardeen had invented the point contact transistor in a way neither Shockley nor Bell had anticipated. When Bell came to patent the point contact transistor in 1948, the patent was granted to Bardeen and Brattain without difficulty. Shockley on the other hand was unable to obtain a primary patent on his field effect principle because of the work carried out before the war by such men as Heil, Pohl and Lilienfield.'[53]

Following the invention of the bipolar transistor at Bell in 1947, Bell Labs continued work on the field effect transistor, with the object of obtaining a device with a higher frequency response than the existing point-contact and grown-junction transistors.

Shockley and Pearson demonstrated the field effect in 1948 and showed that the conductance in the surface region of a semiconductor could be modulated.[54] Following this, in 1953, W.L. Brown postulated the behaviour of an ion-induced conducting channel.[55] In 1955, Ross proposed that such a channel could be made by inducing an inversion layer by means of an electrode placed in the region of the base of a semiconductor device.[56] The difficulty in realising this concept was the lack of a suitable dielectric material across which to induce the charge. This problem was solved by M.M. Atalla in 1959, who proposed that thermally grown silicon dioxide could be used as a gate insulation dielectric, when grown upon single crystal silicon.[57] This was successfully achieved by D. Kahng and M.M. Atalla, of Bell Laboratories, who reported their findings in 1960.[58] This achievement owed much to the work of Ligenza, who succeeded in growing good quality, high dielectric strength silicon dioxide films by high pressure steam oxidation of silicon.[59] Meanwhile in 1957 J.T.Wallmark of RCA took out a patent on the FET but did no further work in this field. By 1959 P. Weiner, also of RCA had constructed a thin-film FET using cadmium sulphide.[60]

The first commercial field effect transistor was produced in France in 1958 by Stanislaus Tesxner, a Polish scientist employed by CFTH, a General Electric affiliate. Called the Technitron, this device was of germanium alloy construction.[61] It suffered from the disadvantages of high reverse leakage current and low transconductance, rendering it effectively a low gain amplifying device.

The first domestic field effect transistor made in the United States was manufactured by Crystalonics in 1960. This device had ten times the transconductance of the earlier Technitron, and was constructed from silicon. Its great advantage was that its noise level was considerably lower than that of the binary junction transistor, particularly at low frequencies. This device, like the Technitron, was of alloy junction construction, and transistors of this type became known as junction field effect transistors (JFET's). In 1962 Texas Instruments introduced the planar junction field effect transistor. This device had a considerably higher transconductance and much lower leakage currents, together with increased production yields and greater uniformity of device characteristics. However, reverse breakdown voltages were lower (about 20 V maximum) which limited it to small-signal applications. Alloy junction devices could sustain much higher voltages before reverse voltage breakdown; for example, Crystalonics marketing a high voltage JFET with a breakdown of 350 V. The silicon FET, upon which a whole new development in the field of integrated circuitry was to rest, was first fabricated by Stevan Hofstein and Frederic Heiman of RCA in 1962.

Other manufacturers now began to produce field effect transistors. By 1962 these companies included Fairchild, RCA, AMELO (a division of Teledyne Inc.) and Siliconix. Cohen writes:

'In the 1962–64 period nearly every semiconductor producer got into the FET business—and out again just as quickly when optimistic predictions failed to materialise.'[62]

A highly important advance made at this time was the invention by the Fairchild Corporation in 1962 of the metal-oxide–silicon field-effect transistor (the MOSFET) using the newly established planar process.[63] This device required fewer steps in its fabrication than its bipolar counterpart and was particularly suited to integrated circuit manufacture, needing only one diffusion compared with, for instance, the bipolar double-diffused epitaxial device which needed four diffusions plus growth of the epitaxial layer. Other advantages were size and weight reduction (for instance, approximately an 18:1 reduction in area) compared with bipolar, furthermore, power dissipation was significantly lower. The main problem arising with the MOS device was that of speed, although Brothers states that:

'at the time (1966) this was not seen as a major disadvantage since over 90% of all logic systems at that time were operating below 1 MHz clock rate (i.e. within the capabilities of MOS).'[64]

Certainly the MOS transistor suffered from being released onto the market in 1964 whilst still suffering from technical problems. These early difficulties resulted in the device being withdrawn. In this context, Brothers writes that:

'The big disadvantage that MOS had in 1966 was its premature release in 1964. The unreliability of the early circuits had soured peoples' attitudes to MOS considerably, even though by 1966 most of the early problems of stability had been cured, difficulties still remained with protecting the circuits from external static fields'.[65]:

In fact the problem of oxide breakdown due to electrostatic charge tended to plague the manufacture of these devices over the next decade and this factor led to their relatively slow acceptance.

Nevertheless, the MOS device, in its integrated circuit form, became a viable product by 1970 largely due to the efforts of two small American firms, General Microelectronics and General Instrument, and during the next decade it was to lead to a complete transformation of the electronic industry, progressively shifting the emphasis from bipolar to MOS digital circuitry, the latter technology possessing the advantages of higher packing density per slice and considerably lower power dissipation.[66] Since fewer process stages were needed, yields were higher and manufacturing costs correspondingly less. A major cause of failure with discrete devices is due to faulty intermetallic connections. Since, in the case of integrated circuitry, electrical connections could be made from one device to the other by aluminum deposition onto the surface of the slice, using photomasking and etching techniques, the need for intermetallic connections was confined to attaching the chip to its external electrical contacts, thus greatly improving reliability. The combination of these two factors—increased reliability with decreased cost, achievable through the use of MOS technology—are the major reasons for the success of the integrated circuit in its present form.

Two basic types of MOS circuit have been fabricated, P–MOS and N–MOS. P–MOS devices are manufactured with an N-substrate and P-type source and drain diffusions. N–MOS devices are manufactured with a P-type substrate and N-type source and drain diffusions. Electron mobility in silicon is about twice that of hole mobility, consequently the N-MOS FET (in which the majority carriers flowing in the channel are electrons) is a higher speed device than its P–MOS counterpart. In addition to its higher speed, the N–MOS FET has the advantage of a higher packing density (a higher number of devices per unit area). However, the N–MOS devices are more difficult to manufacture, owing to the electric charges present at the oxide–silicon interface (often due to contamination) which are positive in polarity and tend to turn on the device. Consequently, most of the earlier devices produced were of the P–MOS type. These difficulties have since been satisfactorily solved and N–MOS has now become the preferred logic for large-scale memories.

An important improvement which took place during the mid-1970s was the development of silicon-gate MOS, which further simplified the fabrication process, allowing smaller devices to be made, with speeds that are now approaching that of bipolar logic. The 'complementary' MOS configuration (CMOS) developed by Wanlass and Sah in 1963 uses both N–MOS and P–MOS devices.[67] This approach has the advantage that power is only taken from the supply during the period when the device actually switches states. Consequently, CMOS has been widely used in such fields as hand held calculators, portable equipment and aerospace systems, in which battery drain is an important factor.

4.11 Genesis of the monolithic integrated circuit

The concept of the monolithic integrated circuit (IC) appears to have first been made public by G.W.A. Dummer of the Royal Radar Establishment (RRE) Laboratories, Malvern (UK). Referring to this event, Roberts writes:

> 'It is widely acknowledged that the first public reference to the possibility of achieving a complete circuit function in a single solid block was made by G.W.A. Dummer at a conference in Washington DC in 1952.'[68]

Roberts states in the same article that:

'I believe that the idea had emerged in discussion between Dummer and members of Caswell Research Laboratory of the Plessey Company Ltd.—specifically G.C. Gout (Director of Research) and N.C. Moore (Research Manager).'

From the outset therefore, a link in this field between a government research establishment and industry was established.

Also, at about this time, the idea that it might be possible to fabricate an integrated circuit had been put forward in the United States and the first patent for such a device was filed on 21st May 1953 by H. Johnson of the Radio Corporation of America (RCA), this being for a 'semiconductor phase shift oscillator and device'.[69] The proposed circuit consisted of a transistor and a resistance–capacitance phase-shift network, the capacitors being reverse-biased P–N junctions. However, it appears that no significant development followed from this proposal.

In the mid-1950s on the initiative of Dummer, a tentative beginning was made in Britain, although according to Atherton:

'Plessey appeared to have regarded the idea as "a laboratory curiosity, or at best as an exploratory feasibility study".'[70]

Kilby states that:

'In 1956, Dummer let a small contract to a British manufacturer (The Plessey Company Ltd). They were unsuccessful in realising a working device primarily because they were working with grown junction transistors.'[71]

However, an appreciation of silicon as a suitable material for the construction of integrated circuits was realised at the time. Golding writes that:

'as early as 1956, researchers in the Physics Group at RRE began to realise that a silicon crystal exhibited all the electrical properties required for the construction of a complete circuit function, an observation which predated the similar in sight of J.S. Kilby at Texas Instruments by some eighteen months.'[72]

As a result of the work carried out at RRE, Dummer mentions that

'a contract was placed with the Plessey Company Research Laboratories in April 1957 for the development of Semiconductor Integrated circuits'[73]

and a model (made of card and paper) was exhibited of an IC flip-flop circuit at the International Components Symposium at RRE in September 1957, this event stimulating considerable American interest. However, according to Roberts, at about this time:

'possible action at Caswell on the true silicon solid-state circuit was delayed by the departure of J.T. Kendall to join Texas Instruments, breaking the continuity of speculation on solid circuits.'[74]

The Plessey team then comprised only five qualified personnel, but were already using advanced techniques, such as diffusion in silicon and etching processes similar to those used later by Kilby. However, in spite of this early start, the integrated circuit concept did not appear to have made any appreciable impact, and when no government funds were forthcoming, the idea was abandoned.[75]

Therefore, during the period when American interest in this field was rapidly growing, little work was being done in the UK despite a promising start, and the early technical lead was lost. Dummer writes that:

> 'Although early work had been done in the field of IC's by Plessey and RRE's Physical Lab., it was not until February 1960 that a team was formed at RRE to study semiconductor techniques.'[76]

At this stage it was felt necessary to obtain Government support in order to stimulate development in the integrated circuit field; consequently a working party was set up by the Admiralty, under the co-ordination of the valve development organisation (CVD) which covered contracts on active (amplifying) devices. The result was that contracts were then placed with a number of companies including Texas Instruments Ltd. (Bedford), The Plessey Company Ltd., Standard Telephones and Cables Ltd., and the Ferranti Company Ltd. The rather leisurely tempo with which developments had proceeded in Britain contrasted with the work carried out in the United States, principally by the Fairchild Corporation and Texas Instruments Ltd (Dallas), the latter firm instituting a six months crash programme in the beginning of 1961, which resulted in the market introduction of a range of planar integrated circuits in August of that year. This programme was instituted following the development of the integrated circuit concept in 1958 by J.S. Kilby, then working for that company.[77] By the summer of the same year he had fabricated the first operational semiconductor circuit, a simple phase-shift oscillator, although using germanium rather than silicon for its construction. He later produced the first working model of a silicon integrated circuit (a simple multivibrator) which was completed on 28th August 1958, this device using silicon pre-etched grown junction transistors. The concept was announced publicly at a press conference in New York on 6th March 1959, and was widely reported in the press.[78] As a result of Kilby's work, the US Air Force issued a contract for further investigation. Within one month from that date Kurt Lehovic of Sprague Company had filed a patent application (22nd April 1959) for the use of reverse-biased P–N junctions to perform the function of isolation diffusion, thus enabling individual components on the chip to be electrically isolated from each other.

The most significant advance was, however, made by R. Noyce of the Fairchild Corporation, who, six months after Kilby's success, produced the first silicon planar integrated circuit, thereby rendering the concept commercially viable.[79] By March 1961, ahead of Texas Instruments, Fairchild produced the first market range of integrated circuits. This method allowed the simultaneous fabrication of transistors, diodes, resistors and capacitors on a single silicon chip. The earliest monolithic integrated circuits were based on 'mesa' technology, and because of the limitations of that technology were inferior to contemporary discrete devices. However, the newly invented planar process was instrumental in providing the desired breakthrough, allowing rapid microminiaturisation to proceed. Improvements in photolithographic techniques quickly reduced the size of the components and enabled multiple circuits to be constructed on a single chip. The rapidity of this development is evident from the following statement made by Brothers:

> 'The complexity of integrated circuits has grown from three transistors and one resistor in 1960 through some 40 odd components in 1964 to some 12 000 components in 1972.'[80]

and in this context Shepherd writes:

> 'thus, in the 1960s bipolar digital circuits progressed from a chip containing a single logical gate to a chip on which a complete small computer central processor can be made.'[81]

This swift and successful development in the United States came about directly as a result of military involvement. Bridges quotes the (US) Director of Defense Research on Engineering stating in a memorandum to the military departments in April 1963: 'This gain in reliability, coupled with reduction in size, weight and power requirements and probable cost savings, make it imperative that we encourage the earliest practicable application of microelectronics to military electronic equipments and systems'.[82] Certainly by the mid 1950s a need had arisen in military circles for a solution to the problem of conflicting demands of increasing complexity versus reliability. In the case of missile systems, the factors of size, weight and power consumption were also vitally important. A further factor was the reduction in maintenance—indeed, in space and missile systems this procedure could hardly be carried out once the system was in operation. The same problems which had presented themselves in connection with thermionic valves towards the latter end of the 1940s, and for which discrete transistors presented a solution, now arose again at a more advanced technical level.

External political factors may also have played a part in the decision of the United States Government to offer substantial assistance to the semiconductor industry at this time. Braun and MacDonald write:

> 'some observers see it as no coincidence that heavy military funding should have commenced so soon after the launching of Sputnik in 1957.'[83]

In that year, a microcircuit programme was started at the Diamond Ordinance Fuze Laboratories, involving printed circuit techniques. (This was a military research organisation which maintained close links with Bell Laboratories.) Important advances were made there at this time in photoresist techniques, and the step and repeat camera, used to design photomasks, was also invented there. These developments were soon to play a vital role in the technology of the planar process.

Several approaches to the problem of circuit integration were attempted; for example, the 'micromodule' supported by the US Army, based on the earlier 'Tinkertoy' concept of discrete devices stacked in a three-dimensional array. Also in the same year (1958) the US Navy sponsored the thin-film approach. Although it was technically feasible to manufacture passive components by this

Table 4.3 US sales of transistors at beginning of 1960s

Year	Germanium Units	Average Value	Silicon Units	Average Value
	× 10⁶	$	× 10⁶	$
1957	27·7	1·85	1·0	17·81
1960	119·1	1·70	8·8	11·27
1963	249·4	0·69	50·6	2·65

Source: EIA Market Data Book, Washington DC, 1974

method, active devices (i.e. transistors) could not be fabricated successfully. A further approach was that of the US Air Force which supported the Westinghouse company in developing what was termed the modular electronics concept, the aim being to construct monolithic units capable of performing a given function without themselves containing discrete elements.

In view of this activity, it certainly appears that at this time the military authorities were urgently working for a solution to the problems arising from the growing complexity of electronic systems and were prepared to fund any approach which seemed to have some chance of success. This climate of opinion could hardly fail to act as a stimulus to those who were attempting to construct integrated circuits, particularly since, should their attempts be successful, the opportunity to supply the highly lucrative military market existed. The intense and rapidly successful efforts at integrated circuit production both by Fairchild and Texas Instruments should be seen in this context.

A further factor acting to encourage semiconductor manufacturers of this time to explore alternatives to discrete devices was that, by the beginning of the 1960s, semiconductor prices had fallen sharply as volume production increased. For example, the US sales of germanium and silicon transistors were as in Table 4.3:

The prospect of declining profits as the market became saturated with discrete transistors cannot be discounted as an additional spur to efforts made by transistor manufacturers, particularly those who specialised in semiconductor fabrication to the exclusion of other activities. In this context, Braun and Macdonald write:

> 'The integrated circuit was, among other things, a solution to the commercial bottleneck problem—too many companies producing too many discrete components at too low prices.'[84]

It would appear, therefore, that an urgent need existed for both the US military authorities and the semiconductor manufacturers to develop the new product, and this was demonstrated by the willingness with which Government and Industry invested considerable sums of money in order to do so. The US Government Deputy Director for Electronics (Department of Defence), J.M. Bridges, writes in December 1964:

'we are sure now that our action in the spring of 1959 was right, that it did advance the availability of usable integrated electronic devices by at least a year, possibly two. In the five years since then, the Department of Defense has spent around thirty million dollars in support of R & D and work on manufacturing methods in integrated electronics. Of much greater significance, however, has been the rapid increase in integrated electronics R & D and production facility support sponsored by industry. It is hard to estimate the amount accurately, but it has certainly been several times the size of the government's effort—perhaps ten times as much'.[85]

The money invested by industry was, however, spent with the knowledge that a substantial military market existed which would absorb the product (in the event of it being successful) at least during the critical initial phase when production yields were low and manufacturing costs high. As can be seen from Table 4.4, initial supplies of integrated circuits were almost entirely to the military during the early years of production. Therefore, as in the case of the discrete transistor, the US Government acted to 'prime the pump' not only financially but by assisting research, development (R & D) and production engineering, but, more importantly, by providing a guaranteed market for the product.

The first commercial application for the integrated circuit in 1963 was in the hearing-aid market,[86] suggesting that the same considerations which determined the original use of the transistor rather than the thermionic valve in this application again applied—for example, a decrease in size and weight and also lower power consumption than its predecessor, the latter factor leading to a decrease in battery costs would tend to compensate for the higher price of the integrated circuit.

Table 4.4 Average uncorrected price of IC's and proportion of production consumed by the military

Date	Average price ($)	% Consumed by military
1962	50·00	100
1963	31·60	94
1964	18·50	85
1965	8·33	72
1966	5·05	53
1967	3·33	43
1968	2·33	37

Source: Tilton J., 'International Diffusion of Technology', 1974 p. 91; and 'EIA Market Data Book', 1975 p. 86

However, the above minor application was to be rapidly dwarfed by the speed with which integrated circuits were to find a major application in providing a solution to the technical and economic problems arising owing to the increased complexity of computers. The decided advantages of integrated circuits over discrete devices, namely increased reliability with decreased cost, together with greater device packing density, ensured an extremely rapid growth in production, primarily supplying military requirements, but shortly after allowing the rapid expansion of the computer into the fields of commerce and industry.

A fundamental distinction between the invention of the transistor and the integrated circuit was that, in the case of the transistor, the initial research and development was primarily a scientific rather than an industrial endeavour, semiconductor manufacturing technology being, at least in its early stages, largely the creation of the thermionic valve companies. Consequently these organisations were faced with the problem of developing entirely new processes and techniques with what was initially a largely unskilled labour force. This was not the case with the integrated circuit. To quote Bridges:

> 'The technology of the semiconductor integrated circuit, however, unlike that of the transistor, grew out of a background of existing solid-state research and marketing expertise. It benefited immediately from the vast research capability that had been built up in the solid-state field, following the first transistor development.'[87]

This factor could hardly fail to assist the development of the integrated circuit, which once under way took place at a relatively rapid rate, aided at the manufacturing level by the existence of a labour force now trained at all levels in the techniques of semiconductor fabrication.

4.12 Influence of large-scale integration upon systems design

A consequence of the move into integrated circuitry has been a dynamic situation in which the demands of ever increasing complexity have conflicted with the need for design flexibility and rapid turn-round time. (Turn-round time is defined as the time taken to design, fabricate and test a given integrated circuit.)

Microminiaturisation has proceeded at a rapid rate because it has the advantage of decreasing weight and size. Less power is dissipated in switching operations, and this has the advantage of reducing power supply requirements; consequentially, less heat is generated to perform a particular operation. Also, as device size decreases, switching time and reliability improve. Furthermore, less silicon per switching element is needed, reducing costs.

However, a problem that arises is that, with increased complexity, design time becomes progressively longer and the technical demands of fabrication become much greater. Consequently, the time taken to construct and test a very large-scale integrated circuit progressively increases as size diminishes. Various approaches have been attempted to overcome this problem, and these will now be discussed.

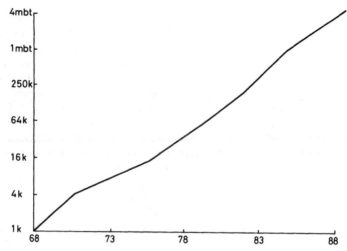

Fig. 4.1 Rate of increase of dynamic read only memory (DRAM)

The development of the integrated circuit industry can, however, hardly be considered without some reference to the parallel growth of the computer industry. A mutually dependent relationship between the two has increasingly determined both the direction of integrated circuit development and the architecture of digital systems. Indeed, computer aided development (CAD) has become an essential component in the production of large-scale integrated circuitry. The major demands of the computer industry have been for low-cost devices possessing high reliability, capable of high operating speed and of minimum cost and weight. A consequent shift towards MOS (and later CMOS) devices has taken place, owing to this technology being more compatible than that of other approaches (for instance, bipolar logic circuitry) with digital system requirements.

An important factor in this development is that in MOS devices, unlike their discrete counterparts, active components such as diodes and transistors tend to be relatively cheap, whilst capacitors and resistors take up much more space on the silicon slice and are consequently more expensive. MOS devices therefore have a particular advantage where large arrays of switching elements are required, as is the case with digital computers. The constant demand by the computer industry for larger memory arrays of ever decreasing size and increasing complexity has led to rapid progress in this area. The number of cells per chip has approximately doubled every year since 1960. This rate of increase has been described as 'Moore's Law' after the American industrialist G.E. Moore.[88] This trend is illustrated by Fig. 4.1.

One result of this increase in chip complexity was that, during the 1960s and early 1970s, basic chip design tended to shift from the area of systems fabrication towards the manufacture of the chips themselves. This in itself led certain systems manufacturers such as IBM to enter chip manufacturing on an in-house basis, but in general the existing demarcations have been preserved and no appreciable shift towards vertical integration has occurred within the American semiconductor industry to date as a result of this trend. Furthermore, because of the increasingly high capital cost in setting up a semiconductor manufacturing facility, any significant move in this direction appears unlikely. It is also important to note that, as logic arrays increase in size, development costs correspondingly increase, this acting as a further deterrent to newcomers intending to enter the device manufacturing field.

A consequence of this rapid increase in the level of complexity is that in general the function of semiconductor chips has tended to become more specific, inhibiting flexibility of system design and also limiting possible market applications. Alternatives to the full custom built circuits have therefore evolved, namely the microprocessor, and a number of what are termed semi-custom circuits. Fig. 4.2, due to Mackintosh, indicates various approaches which have been attempted to satisfy the varied requirements of systems manufacturers.[89]

What now follows is a review of the relative merits of the different approaches to system design with reference to Fig. 4.2

4.12.1 Full custom circuits
This class of integrated circuits include the most complex and large scale circuitry and the semiconductor manufacturer is completely responsible for the design, manufacture and marketing of the product.

Full custom circuits are the original approach to systems design, having directly evolved from the early integrated circuit, and are only distinguished by their greater complexity, being produced by an essentially similar, although technically more advanced, planar process. The advantages of this approach include the ability to achieve maximum packing density and a high degree of design flexibility. A further advantage of this approach is that, because of the high packing density of components on the chip, savings in space and weight may be of significant importance. Also, because of the high degree of circuit complexity, less complex (and therefore less expensive) printed circuitry may be needed. This factor may well offset the relatively higher cost of the full custom chip, particularly since, being a mass produced commodity, unit cost per chip falls rapidly with time as full-scale production develops. The full custom chip

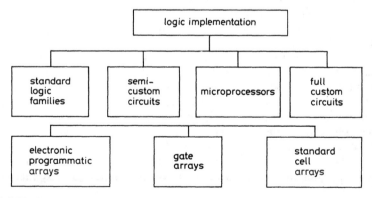

Fig. 4.2 Various approaches to systems manufacture (Source: Reference 89)

also has the advantage that, since fewer intermetallic connections are needed, improved reliability is obtained, and furthermore this advantage becomes increasingly significant with increase in circuit complexity. Unlike semi-custom circuits, design is not limited to a fixed number of pins or to a fixed input/output buffer configuration.

However, because of the complexity of the processes involved, the design and production of this type of circuitry cannot be achieved rapidly and in addition requires considerable technical expertise. Also, in its initial production phase, the full custom chip may be unacceptably expensive, and because of the above disadvantages, alternative approaches to logic design have been devised.

4.12.2 Microprocessors

The microprocessor is essentially the central processing unit (CPU) of a computer fabricated on a single silicon chip. It differs from the 'dedicated' calculator in that it is able to perform a variety of functions determined by the operator by means of programming, thereby giving it considerable flexibility. The calculator has all its operations permanently wired, and circuits must be designed to perform every required function. However, in the microprocessor, because of its programming facility, a variety of possible routes may be used and as a consequence the circuit required is much less. This results in a reduced chip count and also a reduction in weight, space and design effort. Also, because fewer inter-chip connections are required (involving intermetallic contact bonding), reliability is increased.

In order to operate, the microprocessor requires the following additional circuitry:

(*a*) a clock generator in order to control the timing of events constituting the cycle of operation
(*b*) a read/write memory, allowing data storage, entry and recall
(*c*) a read only memory (ROM) in which to store the program instructions
(*d*) input/output devices
(*e*) data address and control lines, which are called buses and used to interconnect the additional circuitry listed above.

This type of control is described as 'software wiring' in contrast to that used by the dedicated calculator which is called 'hardware wiring'. It is software wiring, with its inherent programmable facility, that gives the microprocessor its greatest advantage.

Claims to the invention of the microprocessor have been disputed. For instance, Texas Instruments state that: 'In 1970 Gary Boone of TI developed the first one-chip microprocessor, an integrated circuit having all the elements of a central processing unit (CPU).'[90] However, Braun and Macdonald claim the invention for Ted Hoff, of Intel, who designed the 4004 microprocessor, announced on 15th November 1971.[91]

The invention of the microprocessor was highly significant insofar as by the mid-1970s the emphasis of the semiconductor manufacturing industry was directed away from integrated circuitry towards this new development. The computing power of these devices was rapidly extended. In 1973 Intel augmented the 4 bit 4040 microprocessor with the 8-bit 8080 model. In 1974,

National Semiconductor introduced the 16 bit microprocessor and the 32 bit APX 432 was launched by Intel in 1981.[92] Production of these devices was not confined to a few companies, and by the mid-1970s, virtually all major semiconductor companies in the United States (and also a number in Japan) had entered the microprocessor field.

A disadvantage of the microprocessor compared with the gate array is that of speed. This is because, unlike the gate array, it does not perform only one specific function, and therefore the circuit interconnections on the chip are not optimised to perform at maximum speed of operation.

4.12.3 Semi-custom circuits

This approach provides a method of implementing a particular design in single-chip form, and fabricating it with less delay than would be experienced in producing a full custom circuit.[93] Until the advent of semi-custom logic, the only solution to a parallel logic problem was to use a printed circuit board wired up with individual standard custom chips, or to use a full custom integrated circuit. With the semi-custom approach, the designer is concerned with problems of layout and interconnections, which may be made manually, using similar techniques to those used for printed circuit boards, or, more commonly, computer aided design.

This design technique is economically advantageous for medium-scale production runs, since, if the number produced is too low, the cost of circuit development may not be justified. For very large scale runs, full-custom circuits became a more attractive proposition owing to their more efficient use of silicon, resulting in lower fabrication costs.

Semi-custom circuits may be conveniently classified into (i) electronically programmable arrays, (ii) standard cell arrays and (iii) gate arrays, as suggested by Mackintosh.[94] The latter were first introduced by Texas Instruments in 1965 and are produced as a regular array of logic units on a silicon slice, unconnected to each other.[95] The required interconnecting links between the individual logic elements and also from these elements to the input/output bonding pads are applied at the final stage of fabrication by the technique of aluminum deposition onto the slice surface. This final process may easily be carried out by the user rather than by the device manufacturer since aluminum deposition is a relatively simple matter, involving (at the most) three specific masks. This approach offers the systems manufacturer the advantage of flexibility of design, since changes in circuit configuration involve only changes in masking. Consequently, cost and fabrication time are reduced when compared with the production of standard custom chips. Since wafers are stored in an uncustomised form, production time is typically four weeks shorter than for standard (custom) cells.[96] This method may be used on a large-scale production hire basis and can involve very large-scale integration.

Although the cost of the gate array could be greater than the cost of the small-scale or medium-scale integrated chip which it replaces, savings are likely to be made in printed circuit board manufacturing and also in labour costs, and this factor may render the gate array approach economically viable.

The main disadvantage of gate arrays is that only a single type of logic element is provided and this may result in a waste of silicon area during logic implementation, resulting in turn in an increase in propagation delay and power consumption.

The standard cell design approach to semi-custom design is based on the fabrication by the device manufacturer of a software library of tried and tested cells. This procedure has been followed by several large semiconductor manufacturing companies including NEC and TI. Since standard cells are not fixed as are logic arrays, there is much more flexibility in design and input/output configuration. Standard cell design may be smaller and therefore cheaper than logic arrays in large volume production, although whether this is always the case has been questioned.[97] This approach has flexibility and becomes increasingly important as comprehensive libraries of standard cells are built up by device manufacturers.

Electrically programmable arrays have proved highly useful as a design tool in enabling circuit design to be verified on the slice with minimum capital cost and delay. Making the interconnection patterns on gate arrays involves a process of mask making, metallic deposition and the etching of silicon dioxide. This must be followed by electrical testing, involving in all a period of perhaps more than four weeks before samples are available.[98] To overcome this problem, electrically programmable arrays have been devised. With this approach, the individual cells within the array can be modified without the need of specialist manufacture.

One such type, the erasable programmable read only memory (EPROM) stores information in the form of a capacitance embedded in the insulating silicon dioxide of an individual cell. The memory itself is composed of a large number of these cells connected in matrix form. The application of a voltage pulse to an individual cell charges its associated capacitance, this stored charge constituting one bit of information. This information may be erased by applying ultra-violet radiation to the surface of the cell. The EPROM is a MOS device and relatively slow in operation. It finds an application as a program storage element in microprocessors. Faster programmable read only memories have been made by the bipolar process, but by their nature cannot operate on a charge storage principle. The memory may be programmed by passing a sufficiently high current through the desired cells, causing a metallic element to fuse, thereby producing an open circuit. Once this operation has been performed, however, unlike the EPROM, it is impossible to re-programs the array, except by fusing further elements. More recently, during the last decade, field programmable logic devices have been constructed and, like the above approaches, Bostock points out that development is virtually cost-free and instantaneous, and no minimum quantity commitments are required.[99]

Standard logic families consisting of both bipolar and MOS devices have been developed in the form of integrated circuitry. The most important are as given in Table 4.5.

Logic families may be either bipolar or MOS. Historically, bipolar standard logic families were first in the field, and until about 1975 the U.S. integrated circuit market was broadly based on bipolar transistors.[100] Since then, however, digital MOS integrated circuits have replaced bipolar devices as market leaders and this trend had continued to **gain** momentum. Furthermore, this development has since been repeated both in Europe and Japan.[101]

Table 4.5 Standard logic families (*a*) *Bipolar type*

	Firm	Date announced	
Resistor–Transistor logic (RTL)	Fairchild	1961	Largely
Resistor-coupled transistor logic (RCTL)	Texas Inst.	1961	Obsolete
Diode–Transistor logic (DTL)	Signetics	1962	by mid-1970s
Transistor–Transistor logic (TTL)	Sylvania	1964	In
Emitter–coupled logic (ECL)	Motorola	1963	current
Integrated injection logic (I^2L)	Philips	1968	use

(*b*) *Field effect type*

Metal-oxide–silicon (MOS)	General Micro Electronics	1965

Source: GOLDING and others

The three earliest logic families listed in Table 4.5 (RTL, RCTL and DTL) were originally manufactured as discrete devices and only produced in integrated circuit form at the stated date. The three later bipolar logic families mentioned were designed specifically as integrated circuits and are currently in production. TTL, in particular, has been extremely successful and several variants of this type have since been produced. ECL is faster than TTL, but has the disadvantage of higher power consumption. Both are faster in operation than MOS type devices, although the gap has narrowed considerably during recent years. Sze, writing in 1983, mentions that, owing to decrease in device dimensions, switching time in MOS devices has improved by two orders of magnitude since the original devices were marketed.[102] This factor has certainly been significant in accelerating the trend from bipolar to MOS technology. A further factor has been that, owing to its structural simplicity compared with bipolar devices, a far greater degree of microminiaturisation has been possible,- leading to a progressive decrease in power consumption per bit (this being the lowest of all in the case of CMOS) and therefore less energy is needed to perform a switching operation. Nevertheless, a considerable market for standard TTL and ECL chips continues to exist, offering as they do design flexibility and speed of operation.

4.13 Small-signal semiconductor diode

The germanium point-contact diode originated from wartime work in the ultra-high-frequency radar field. Great advances were made in improving diode characteristics during this time, much important work being done at Purdue University. By October 1943 it had become possible to fabricate detectors with reverse voltage breakdowns of 150 V, using polycrystalline silicon.[103]

In 1952 Transitron started to manufacture the gold bonded diode.[104] This is a point-contact device in which a gold whisker, doped with a small percentage of gallium or indium, is brought into mechanical contact with the N-type germanium pellet. When a current pulse is passed, alloying occurs, the gallium or indium forming a P–N junction. The characteristics of this device are generally superior to those of the germanium point-contact diode, with the exception of frequency response, which is somewhat lower, owing to the gold bonded diode having a rather higher depletion layer capacitance. Nevertheless, in spite of the existence of the gold-bonded diode, most small signal diodes manufactured in the 1950s were of the germanium point contact type, the characteristics of which could be varied by adjusting the doping level of the germanium. For example, by this means it was possible (by lightly doping) to fabricate devices with breakdown voltages up to about 200 V, with a frequency cut-off of about 5 MHz. Alternatively, it was possible (by increased doping) to produce diodes with breakdown voltages of about 25 V, but with a frequency cut-off of several hundred megahertz.[105]

By about 1960, silicon diodes were beginning to replace the earlier germanium type, this change being due to their superior performance over a wide range of parameters.

Silicon alloy units were first constructed at Bell Laboratories by Pearson and Sawyer in 1952, the process consisting of alloying an aluminum (P-type) wire onto an N-type silicon chip.[106] These devices possessed the advantage over the germanium alloy diode of having a higher voltage rating, a higher current capacity, lower reverse leakage currents and a higher temperature rating. One disadvantage, however, was that they generated reverse current spikes at high frequencies. Principally owing to their high temperature characteristics, they were in demand by the military.

A further development in this field was the silicon diffused junction rectifier, described by Prince in 1956.[107] These diodes had peak inverse voltages of up to 400 V with current ratings of 400 mA. This construction was particularly rugged and reliable, and consequently also generally in demand by the military. Following the invention of planar technology in 1962, planar diffused junction rectifiers were manufactured and these exhibited further reductions in reverse leakage current, and improved breakdown voltage characteristics. They are described in a paper by Olson and Robilliard, published in 1967.[108]

4.14 Power rectifier

Semiconductor diodes cover a very wide spectrum of operation. Hibberd, writing in 1959, mentions a range comprising at the one extreme crystal diodes, operating at a few microwatts at 30 GHz, and at the other end of the scale, silicon power rectifiers handling up to 20 kW at 50 Hz.[109] Given this fact, it is difficult to decide exactly how to define a power diode. One suggestion is that a semiconductor diode becomes a power diode when mounted on a heat sink, since this action implies a desire to dissipate power.[110] Under these conditions, the device is generally referred to as a power rectifier, since its principal use is that of rectification.

The development of power diodes has led to considerable savings in weight, size and cost of electrical systems, and they also have the advantage of ruggedness when compared with their mercury-arc predecessors. For example, Bates and Colyer mention that the semiconductor diode was a very convenient replacement for arc devices such as ignitrons, its smaller size and fewer auxiliaries allowing considerable savings to be made in equipment size and weight, this being particularly important for portable equipment.[111] They also note that their size and robustness led to them being built into the rotating parts of electrical machines, leading to the brushless alternator.[112]

Progress in the development of the power rectifier was rapid during the 1950s. Roualt and Hall describe a power rectifier of the germanium alloy type in 1952, and devices handling up to 10 A were available in that year.[113] By 1953, germanium alloy power rectifiers were being manufactured which could handle 50 A. However, it became apparent that, despite the lower forward voltage breakdown characteristic of germanium when compared with silicon, the limiting factor on current handling capacity was operating temperature. Here silicon had the advantage, since germanium was limited, owing to thermal runaway, to a maximum operating temperature of 75 to 80xC, compared with a value of about 175xC for silicon. With water cooling, it was possible to design germanium rectifiers to handle 200 A at a peak inverse voltage of 600 V.[114] Water cooling, however, limited the advantages of the device and consequently opinion in the industry moved in favour of the silicon rectifier, which was also being rapidly developed. As early as 1952, Bell Laboratories had produced an alloy junction silicon unit and in 1955 General Electric began to manufacture silicon power devices having a greatly improved reverse voltage characteristic when compared with germanium. By 1959, silicon power rectifiers capable of handling 100 A with reverse breakdown voltages of 300 V were being made, and by 1969 single units which could handle 125 A and with reverse breakdown voltages of 1200 V were coming onto the market.

4.15 Thyristor

Thyristors, or silicon controlled rectifiers, are semiconductor devices with three junctions and a P–N–P–N structure, which not only rectify but allow control of current flow by a signal applied to the 'gate' terminal. By this means, large currents can be controlled by a very small signal. Essentially, therefore, the device acts as a controlled switch.

Shepherd states that this device was developed in silicon as an extension of the original P–N–P–N structure described by Moll in 1956.[115,116] From that time onwards, progress in the development of the thyristor was quite rapid. In the early 1960s silicon thyristors capable of handling up to about 100 A and

300 V were introduced and by 1970 a single unit could handle almost 1000 A and withstand 2500 V.[117]

These devices made the small-scale conversion of alternating current to direct current feasible and economic. They are much more efficient than mercury-arc devices, including the grid-controlled mercury arc rectifier, which they have replaced (the high voltage drop in the mercury arc making this type of device inefficient at low voltages). By comparison, thyristors are simpler, smaller and more suitable for low voltage applications, and they have excellent reliability, provided that they are protected against uncontrolled current and voltage surges. They find wide application as power control units, supplanting the thyratron valve. Bates and Colyer point out that the impact of the thyristor was much greater than its mercury arc predecessor, not only for the reasons previously stated, but because even from its early days it was a readily obtainable device, and because of this factor much more widely used in industrial laboratories and universities.[118] The likely result of this would be to increase the range of its possible applications.

Unlike the small-signal transistor, the thyristor is a relatively high cost device, and numbers produced are relatively low. Since they are largely used in power engineering, controlling very large currents and voltages, reliability is an important factor. There is therefore a significant testing element which is reflected in the device unit cost. Thyristors, because of their specialised applications, are manufactured by only a limited number of companies and these tend to be connected with the electrical power industry.

Specific applications of silicon controlled rectifiers lie in the field of speed and power control in the steel industry, and also in paper and textile mills. More recently, they have found numerous uses in the domestic market, including sewing machines, slide projectors, washers and dryers, and also in heating and ventilating systems.

Fig. 4.3 shows a steady increase in thyristor sales in the United States in terms of dollar values over the 1960s and 1970s. Although this increase has not since been maintained it should be remembered that, also during this period,

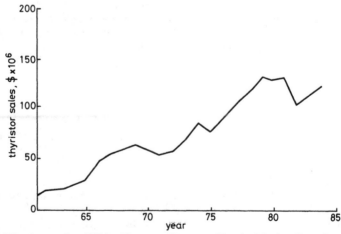

Fig. 4.3 Thyristor sales (USA) (Source: *Electronics* Yearly Market Data Survey)

unit values have fallen substantially. The actual rate of increase in units produced during this time is therefore significantly higher than that indicated by Figure 4.3.

4.16 References

1 BRAUN, E. and MACDONALD, S.: 'Revolution in miniature' p. 45
2 WEINER, C.: 'How the transistor emerged' *IEEE Spectrum*, Jan. 1973, p. 28
3 *Ibid.*, p. 31
4 KELLY, M.: *Proc. Royal Soc.* (Series A), 1950, **203**a, pp. 287–301
5 GOLDING, A.M.: D.Phil Thesis, University of Sussex, 1971, p. 327
6 SHOCKLEY, W.: '25 years of transistors', *Bell Labs. Record*, Dec. 1972, pp. 340–341
7 WEINER, C.: *op.cit.*, p. 32
8 SHOCKLEY, W.: 'The theory of P–N junctions in semiconductors and P–N junction transistors' *Bell System Tech. J.* 1948, **28**, p. 435
9 SHOCKLEY, W.: 'The path to the conception of the junction transistor' *IEEE Trans.*, 1976, **ED–23**, p. 599
10 TEAL, G.K.: 'Single crystals of Ge and Si basic to the transistor and integrated circuit' *IEEE Trans.*, 1976, **ED–23**, p. 624
11 *Ibid.* p. 628
12 O'CONNOR, J.: 'What can transistors do?' *Chemical Engineering*, 1952, **59**, p. 156
13 SPARKES, M.: 'The junction transistor' *Scientific American*, July 1952, p. 32
14 SHOCKLEY, W.: *IEEE Trans.*, 1976, **ED–23**, p. 597
15 HIBBERD, R.G.: *Proceedings IEE*, Part III May 1959, pp. 264–278
16 OBERCHAIN, I.R. and GALLOWAY, W.J.: 'Transistors and the military' *Proceedings IRE*, 1952, **40**, pp. 1287–1288
17 BELLO, F.: 'The year of the transistor', *Fortune*, March 1953, p. 128ff
18 STEUTZER, O.M.: *Proceedings of the IRE*, 1952, **40**, pp. 1529–1536
19 HALL, R.N. and DUNLAP, W.C.: 'P–N junctions prepared by impurity diffusion' *Physics Rev.*, 1950, **80**, pp. 467–488
20 LAW, R.R., MUELLER, C.W., PANKOVE, J.J. and ARMSTRONG, L.O.: 'A developmental germanium P–N–P junction transistor' *Proceedings IRE*, 1952, **40**, pp. 1352–1357
21 SABY J.S.: 'Fused impurity P–N junction transistors' *Proceedings IRE*, 1962, **40**, pp. 1358–1360
22 SHEPHERD, A.A.: 'The properties of semiconductor devices', *J. Brit. IRE*, 1957, **17**, pp. 262–273
23 MORTON, H.L.: 'The present status of transistor development', *Bell System Tech. J.* May 1952, p. 435
24 SHEPHERD, A.A.: 'The properties of semiconductor devices', *op.cit.* p. 268
25 TEAL, G.K.: 'Some recent developments in silicon and germanium materials and devices'. Presented at the National Conference on Airborne Electronics, Dayton, Ohio, 10th May 1954
26 ADCOCK, W.A., JONES, M.E., THORNHILL, J.W. and JACKSON, E.D.: 'Silicon transistor', *Proc. IRE*, 1954, **42**, p. 1192
27 TEAL, G.: 'Announcing the transistor', *25th Anniversary Observance Transistor Radio and Silicon Transistor*, TI Inc., 1980, p. 1
28 *Ibid.*
29 GOLDING, A.M. *op.cit.* p. 158
30 SANGSTER, R.: 'Announcing the transistor', *25th Anniversary Observance Transistor Radio and Silicon Transistor*, TI Inc., 1980, pp. 1–2
31 ADCOCK, W.: *op.cit.*
32 FROSCH, C.J. and DERRICK, L.: 'Surface protection and selective masking during diffusion in silicon', *J. Electrochemical Soc.*, 1957, **104**, p. 547
33 ASCHNER, J.F., BITTMAN, C.A., HARE, W.F.J. and KLEINACH, J.J.: 'A double diffused silicon high frequency transistor produced by oxide masking techniques', *J. Electrochemical Soc.*, 1959, **106**, p. 1145–7
34 SPARKES, J.J.: 'The first decade of transistor development', *Radio & Electron Eng.*, 1973, **43**, p. 8

35 GRINICH, V.H. and HOERNI J.A.: 'The planar transistor family', *Colloque International sur les dispositifs semiconductors*, Paris, Feb. 1961
36 SPARKES, J.J.: 'The first decade of transistor development', *Radio & Electron. Eng.*, 1973, **43**, p. 8–9
37 DUMMER, G.W.A.: 'Electronic inventions 1745–1976' p. 126
38 *Ibid.*, p. 107
39 SHEPHERD, A.A.: 'Semiconductor developments in the 1960s', *Radio & Electron. Eng.*, 1973, **43**, p.15
40 ROKA, E.G., BUCK, R.E. and REILAND, G.W.: 'Developmental germanium power transistors' *Proc. IRE*, 1954, **42** pp. 1247–1250
41 HIBBERD, R.G.: 'Transistors and associated semiconductor devices', *IEE*, Paper 2914, May 1959
42 SHEPHERD, A.A.: *Ibid.* p. 16
43 OVERTON, B.R.: 'Transistors in Television Receivers', *J. Television Soc.*, 1958, (8), p. 444
44 MORTON, J.A.: *Bell Syst. Tech. J.*, May 1952, p. 435
45 SHEPHERD, A.A.: 'The properties of semiconductor devices', *J. Brit. IRE*, 1957, **17**, p. 265
46 HIBBERD, R.G.: 'Transistors and associated semiconductor devices', *IEE*, Paper 2914, May 1959
47 THORNTON, C.G. and ANGELL, J.B.: 'Technology of micro-alloy diffused transistor', *Proc. IRE*, June 1958, p. 1166
48 BEALE, J.R.A.: 'Alloy diffusion: A process for making diffused base junction transistors', *Proc. Phys. Soc. B*, 1957, **70**, p. 1087
49 LEE, C.A.: 'A very high frequency diffused base germanium transistor', *Bell Syst. Tech. J.*, 1956, **35**, p. 23
50 BELLO, F.: *op.cit.* p. 132
51 'Reducing internal dimensions improves silicon transistor performance', *Bell Labs. Record*, 1967, **45**, pp. 13–15
52 MORRIS, P. R.: 'Historical survey of early research in semiconductors', M.Phil thesis, pp. 37–39
53 BRAUN, E. and MACDONALD, S.: *op.cit.* p. 40
54 SHOCKLEY, W. and PEARSON, G.L.: *Phys. Rev.*, 1948, **74**, p. 232
55 BROWN, W.L.: *Phys. Rev.*, 1953, **91**, p. 518
56 ROSS, I.M.: US Pat. No. 2 791 700 (filed 1955; issued 1957)
57 ATALLA, M.M.: US Pat. No. 3 206 670
58 KAHNG, D. and ATALLA, M.M.: 'Silicon–silicon-dioxide field induced surface devices'. IRE–AIEE Solid State Device Res. Conference, Carnegie Institute of Technology, Pittsburgh, PA 1960
59 LIGENZA, J.L. and SPETZNER, W.G.: *J. Phys. Chem. Solids*, 1960, **14**, p. 131
60 ANTEBI, E.: 'The electronic epoch', (Van Nostrand, 1982)
61 COHEN, J.M.: 'An old timer comes of age', *Electronics*, 19 February 1968, p. 123
62 COHEN, J.M.: *Ibid.*
63 *Ibid.*
64 BROTHERS, J.S.: 'Integrated circuit development', *Radio & Electron. Eng.*, 1973, **43**, p. 43
65 *Ibid.*
66 BRAUN, E. and MACDONALD, S.: *op.cit.* p. 102
67 WANLASS, F.M. and SAH, C.T.: 'Nanowatt logic using field effect metal oxide semiconductor triodes' *ISSCC Digest*, Feb. 1963, p. 32–33 US Patent 3 356 858 (Filed June 1963; issued 1967)
68 ROBERTS, D.H.: 'Silicon integrated circuits' *Electron. & Power*, April 1984, p. 282
69 JOHNSON, H.: US Patent No. 2816228, 10 December 1957 (Filed 21st May 1963)
70 ATHERTON, A.A. 'From compass to computer, p. 251–252
71 KILBY, J.S.: *IEEE Trans.*, **ED-23**, pp. 648–654
72 GOLDING, A.M.: *op.cit.* p. 353
73 DUMMER, G.W.A.: 'Integrated electronics development in the UK and W. Europe', *Proc. IEEE*, Dec. 1964, p. 1415
74 ROBERTS, D.H.: *Ibid.*
75 Conversation with G.W.A. Dummer, May 1985

76 DUMMER, G.W.A.: *Ibid.*
77 KILBY, J.S.: US Patent No. 3 138 743 (Filed 6th February 1959). Also Patent No. 3 261 081 (J.S. Kilby and Texas Instruments) (Filed 19th July 1966)
78 KILBY, J.S.: *IEEE Trans.*, 1976, **ED–23**, pp. 648–654
79 NOYCE, R.N. US Patent No. 2 981 877 (Filed 30th July 1959)
80 BROTHERS, J.S.: 'Integrated circuit development', *Radio & Electron. Eng.*, 1973, **43**, p. 39
81 SHEPHERD, A.A.: 'Semiconductor device development in the 1960s' *Radio & Electron. Eng.*, 1973, **43**, p. 17
82 BRIDGES, J.M.: 'Integrated electronics in defence systems' *Proc. IEEE*, Dec. 1964, p. 1407
83 BRAUN, E. and MACDONALD, S.: *op.cit.* p. 95
84 *Ibid.*, p. 88
85 BRIDGES, J.M.: *op.cit.*
86 GREENBAUN, W.H.: 'Miniature radio amplifier' *J. Audio Eng. Soc.*, 1967, **15**, p. 442
87 BRIDGES, J.M.: *op.cit.*
88 MOORE, G.E.: Fairchild, and later from 1968 Vice President of Intel
89 MACKINSTOSH, I.M.: 'Profile of the World–wide Semiconductor Industry', 1982, p. 118
90 TI: 'The microprocessor, the device of the future', TI Publication, Bedford
91 BRAUN, E. and MACDONALD, S.: *op.cit.* p. 108
92 *Ibid.* p. 112
93 TOBIAS, J.R.: 'LSI/VLSI building blocks', *IEEE Computer*, 1981, **14**, p. 83–101
94 MACKINTOSH, I.M.: *op.cit.* p. 118
95 *Ibid.* p. 117
96 Hitachi Publication: 'Gate arrays CMOS & BiCMOS from concept to silicon'
97 HEATON, R.: 'Some user notes on semi-custom logic' *in* READ, J.W. (Ed.): 'Gate arrays' (Collins, 1985) p.313
98 BOSTOCK, G.: 'Bipolar and high speed arrays', *Ibid.*, p. 64
99 *Ibid.*, p. 69
100 SZE, S.M.: 'VLSI technology' (McGraw Hill) p. 1
101 For a detailed analysis, see MACKINTOSH: 'Profile of the world-wide semiconductor industry', 1986; also *'Electronics'* market forecasts and trends', yearly
102 SZE, S.M.: 'VLSI Technology' (McGraw Hill) p. 4
103 BRAUN, E. and MACDONALD, S.: *op.cit.* p. 29
104 HIBBERD, R.G.: 'Transistors and associated devices', IEE Paper 2914, May 1959
105 *Ibid.*
106 PEARSON, G.L. and SAWYER, B.: 'Silicon P–N junction alloy diodes', *Proc. IRE*, 1952, **40**, pp. 1348–1351
107 PRINCE, M.B.: 'Diffused P–N junction silicon rectifiers', *Bell Syst. Tech. J.*, 1955, **35**, p. 661–84
108 OLSON, K.H. and ROBILLARD T.R.: 'New miniature glass diodes', *Bell Laboratories Record*, 1967, **45**, pp. 13–15
109 HIBBERD, R.C.: 'Transistors and Associated Semiconductor Devices', IEE Paper 2914, May 1959, p. 269
110 Conversation with T.G. Brown (Mullard Ltd.)
111 BATES, J.J. and COLYER, R.E.: 'Electrical power engineering', *Radio & Electron. Eng.*, 1973, **43**, pp. 116–117
112 FORD, A.W.: 'Brushless generator for aircraft—a review of current developments', *Proc. IEE*, 1962, **109**, pp. 437–55
113 ROUALT, C.L. and HALL, G.N.: 'A high voltage, medium power rectifier' *Proc. IRE*, 1952, **40**, pp. 1519–1521
114 SHEPHERD, A.A.: 'Semiconductor device developments in the 1960s', *Radio & Electron. Eng.*, 1973, **43**, p. 13
115 SHEPHERD, A.A.: *Ibid.* p. 14
116 MOLL, J.L., TANENBAUM, M., GOLDNEY, J.M. and HOLONYAK, N.: 'P–N–P–N transistor switches', *Proc. IRE*, 1956, **44**, pp. 1174–1182
117 SHEPHERD, A.A.: *op.cit.* p. 14
118 BATES, J.J. and COLYER, R.E.: 'Electrical power engineering', *Radio & Electron. Eng.*, 1973, **43**, p. 116

Chapter 5

Major technical processes used in semiconductor device fabrication

The following chapter contains a brief review of the technical processes essential to the fabrication of semiconductor devices. It is not intended as a technical treatise, but rather to give an outline of these major processes and consider their development within a historical framework, thereby complementing the survey in chapters 6, 7 and 8, which deal with the major non-technical aspects of the industry's growth. It should be remembered, however, that both technical and non-technical aspects of this growth are interrelated, and that this division is merely one of convenience.

5.1 Preparation of high purity semiconductor material

It is doubtful if any significant progress could have been made in the development of the transistor without the ability to produce sufficient quantities of high purity semiconductor material. A highly important advance in this respect was the process of zone refining, developed by W.G. Pfann, of Bell Laboratories, in 1951; the basic paper of zone refining being published in 1952.[1] This process was vital because for the first time it became possible, by a relatively simple technique, to produce the materials of extremely high purity needed for semiconductor manufacture, and from which high quality semiconductor crystals could be grown. The history of this development is somewhat curious. W.G. Pfann writes that:

> 'Several years after my paper had appeared I learned that Peter Kapitza, the Russian physicist, had actively performed a one-pass zone melting operation and had described it in a paper published in 1928. He was concerned with crystal growth, however, and apparently did not recognise the potential of the molten zone as a distributor of solutes and it would seem that neither did any of the hundreds of physicists who must have read his paper.'[2]

Pfann, when working at Bell Laboratories in 1939, was allowed to spend half his time on fundamental research of his own choosing and during this period invented the process of zone levelling; i.e. he grew lead crystals by passing a short melted zone along a lead ingot 'levelling out' the distribution of the major impurity present (antimony). He comments that:

'this procedure, now called zone levelling, did not strike me as particularly remarkable, and nothing came of it at the time. I assumed that such a simple idea must be common knowledge'.[3]

Soon Pfann became engaged in wartime problems, and did no further work in this field until much later when, as he states, 'the need arose for geranium of uniform purity for transistors'. It was at this stage when, as he writes:

'Suddenly the lightning of zone refining struck: a molten zone moving through an ingot could do more than level out impurities—it could remove them.'[4]

He showed that by means of successive 'passes' of the molten zone through the ingot it was possible to reduce unwanted impurities to an extremely low level, typically less than ten atoms per cubic centimetre. This occurs because most impurities found in semiconductor materials such as germanium and silicon have a lower melting point than the parent material and tend to stay in the molten zone rather than freeze out; consequently the impurities are swept into the end of the semiconductor ingot, where they can be physically removed by lopping off that portion of the bar.

In spite of Pfann's vital contribution to transistor technology he received little initial encouragement. Braun and MacDonald write that Pfann was:

'actively dissuaded from pursuing his ideas. Shockley, among others, did not think that zone refining would be important in the semiconductor field.'[5]

Although the process of zone refining silicon is more difficult, nevertheless it has been possible to use this process to purify that element to approximately the same degree as in the case of germanium. This has been achieved by a modification of the zone refining process known as the float zone method, invented by H.C. Theurer (a colleague of W.G. Pfann) in 1953. The process was independently discovered by T.H. Keck and M.J.C. Golay of the US Army Signal Corps Laboratories at Fort Monmouth, NJ (who published first, and gave the technique its name) and by R. Emies of Siemens & Halske. This method uses the surface tension effect to support a stable liquid zone formed by induction heating, the silicon rod being mounted vertically in a reducing atmosphere rather than being held in a crucible, thus avoiding contamination.

5.2 Etching techniques

Etching processes have been widely used in the fabrication of all types of semiconductors. Specifically in the planar process, upon which all subsequent device development rests, etching is used to open 'windows' in silicon dioxide in order to diffuse impurities into the silicon slice and also to remove unwanted aluminum during the contact deposition process. The original approach to window etching was to mark the area of oxide to be protected, using a photoresist, and remove the oxide from the unprotected area with hydrofluoric acid (HF), a selective etch which would not attack the underlying silicon, thus exposing the silicon surface for purposes of diffusion. Although this 'wet' etching technique was successful with discrete devices and small-scale integration, difficulties arose when dealing with smaller geometries, owing to lateral 'undercutting' of the photoresist, which thereby introduced a significant error in the geometry of the diffused area. Various techniques have been evolved in overcoming this problem and may be described under the heading of 'dry etching' or 'plasma-assisted etching'.

The earliest application for plasma etching in connection with integrated circuits dates from the late 1960s when techniques of photoresist stripping using oxygen plasmas were being investigated.[6] Work was also being carried out at this time on the use of plasmas for etching silicon and this is described in a patent filed by Irving, Lemons and Bobos in March 1969.[7] In the early 1970s investigations were being made in connection with the deposition of silicon nitride upon silicon, the purpose being to produce an electrically stable passivating surface layer. However, no suitable wet etchant could be found to etch windows in silicon nitride. This problem was however solved by the use of plasma etching,[8] and this work marked the first significant application of this technique in integrated circuit manufacture. It was later realised that plasma etching exhibited anisotropic properties, namely, the vertical etching rate considerably exceeded the lateral etching rate, making this approach particularly suitable for work involving very small geometries. By the mid-1970s most major integrated circuit manufacturers were adopting this method as a preferable alternative to the 'wet' etching process.

5.3 Diffusion process

A most significant advance in the preparation of P–N junctions was that of diffusion from the vapour phase of opposite types of impurity into a semiconductor, described by Fuller and Ditzenburger in 1956.[9] Petritz states that:

> 'research on the diffusion of III–V impurities into germanium and silicon by Fuller at BTL and by Dunlap at GE laid the foundation for transistor fabrication using diffusion as a key process step. The BTL was first to fully integrate these results into germanium and silicon transistors.'[10,11,12]

The great advantage of the diffusion process is that the doping impurities diffuse at a very slow rate which can be controlled by adjusting the temperature at which diffusion takes place. Consequently, junction depth can be controlled to very fine limits and results are readily reproducible. The resulting junction is of the graded type. Because of this very fine control of junction depth, devices with very narrow base-widths may be manufactured, permitting the construction of transistors with very high frequency characteristics. Using this technique, the upper limit of frequency response has been extended to hundreds of

MHz. Furthermore, the graded base permitted a high base-collector breakdown voltage to be achieved together with a narrow basewidth, a combination of desirable characteristics previously unrealisable.

By the mid 1960s the diffusion method had almost completely supplanted other types of transistor junction fabrication and was an essential ingredient in the construction of the mesa and planar transistor. However, diffusion was first used in conjunction with already existing methods of manufacture, and both the alloy and grown junction processes were modified to incorporate the advantages offered by this new technique. Petritz describes a germanium diffused-base alloyed-emitter type device, in which the base–collector junction had been etched in order to improve the base–collector reverse breakdown characteristics.[13] This etching process resulted in a table-like profile in shape not unlike the 'mesa' type mountains in the southern United States, and consequently devices etched in this fashion became known as 'mesa' devices. Photomasking techniques began to be used at this time in conjunction with etching in order to produce the required mesa junction surface profile.

The Philco jet-etching technique was also modified to produce the micro-alloy diffused transistor (the MADT) described by Thornton and Angell, the N-type germanium base region being diffused prior to jet etching and plating of contacts, this resulting in an improvement in frequency response, coupled with low noise properties.[14]

All the devices mentioned so far—with the exception of the planar transistor—suffered from long term instability owing to the unpassivated state of their junction surfaces. Leakage currents were relatively high and increased with temperature, necessitating specially designed direct current biasing circuits to reduce these effects. Because of the complexity of surface states, a theoretical approach was difficult; therefore efforts to produce electrically stable characteristics tended to lead to an empirical approach on the part of engineers engaged in the production of these devices. It became increasingly obvious towards the latter end of the decade under consideration that what was needed was a means of electrically stabilising the surface of the transistor. The achievement of surface stabilisation by the deposition of an electrically stable layer of silicon dioxide covering the junctions, together with the technique of diffusion, was to lead to a qualitative advance in semiconductor technology, namely, the silicon planar process.

With the advent of the planar transistor, the diffusion process was universally used, enabling extremely accurate control of junction depths to be achieved. (A typical diffusion rate being one micron per half an hour.) Consequently, fine control of base width could be obtained, resulting in the ability to produce devices with improved high frequency response. Furthermore, advances in temperature control of diffusion furnaces during the mid 1960s enabled larger batches of silicon slices than previously to be processed, with extremely uniform electrical characteristics, increasing production throughout and decreasing unit costs.

Recent advantages in production technology, involving more complex circuitry and decreasing device size, notably in the integrated circuit field, have imposed more stringent requirements upon not only this process but also upon methods of measurement of diffusion profiles.

5.4 Surface stabilisation of silicon by the oxide masking process

The technique of depositing a layer of oxide onto the surface of silicon, thus electrically stabilising it, was first developed by C.J. Frosch of Bell Laboratories in 1957.[15] The oxide so formed, silicon dioxide, is a glass and therefore an electrical insulator, and impervious to moisture.

Frosch showed that a layer of thermally grown silicon dioxide prevented the penetration of elements commonly used as dopants during the diffusion process, and could therefore be used as a mask to prevent selected areas of the silicon surface from these diffusants. In addition, when used as a protective barrier covering the P–N junction, it effectively prevented surface contamination both during the remainder of the manufacturing process and thereafter. An important consequence of this development was the reduction of reverse leakage currents, increase in reverse breakdown voltages and a significant improvement in long-term stability of the device.

This method of surface passivation was to have highly important consequences in making possible an entirely new method of transistor manufacture, the planar process, upon which the subsequent development of the industry has rested. In addition, it has been a significant factor in enabling the integrated circuit to be successfully fabricated. One important result was that this process firmly established silicon as the prime material for transistor and integrated circuit manufacture, since it is not possible to grow a comparable electrically stable oxide onto a germanium surface.

Recent demands in the field of integrated circuitry have imposed additional emphasis upon the ability to grow thermally stable oxides. These requirements range from highly reliable thin oxides to thick isolation oxides that can be grown at moderate temperatures.

Also, with the advent of integrated circuits, applications of the oxidation process have been extended to cover not only masking against diffusion of the dopant into silicon and the provision of surface passivation, but, in addition, dielectric isolation, used as a component in metal-oxide–silicon structures, and to provide a means of electrical isolation of multilevel metallisation systems.

5.5 Epitaxial process

Although diffusion may be used to produce low resistivity layers on a high resistivity substrate, it cannot produce high resistivity layers on a low resistivity substrate. However, the process of epitaxy may be used to obtain such a resistivity profile.

In June 1960 Bell Laboratories announced a method of growing single crystal material from the vapour phase, the material under investigation being silicon.[16] Work had, however, been carried out on this process at Bell as early as 1950, in connection with germanium.

The epitaxial layer is deposited on a substrate of the same material and takes up the same crystal orientation as that presented at its surface by that material. Although this high quality epitaxial layer may be formed by this process, crystal dislocations and imperfections tend to increase as the epitaxial layer increases in thickness and imposes a practical limit upon this dimension. Impurity doping can be carried out during epitaxial deposition and the impurities deposited may be of the same or opposite type to that of the parent substrate.

In semiconductor manufacture the device is diffused into the epitaxial film layer, the substrate being unaffected. Consequently it is able to give mechanical strength to the device without adversely affecting its electrical characteristics. For example, the layer resistivity may be optimised to yield a high-reverse–breakdown collector and base of a transistor whilst the substrate may be of low resistivity, thus reducing the collector series resistance. A high collector series resistance is undesirable, firstly, because it causes heat to be dissipated within the material, raising its temperature and increasing thermal leakage currents, secondly, because it is a source of power loss, and, thirdly, it may form a time constant in conjunction with the base–collector depletion layer, adversely affecting the switching properties of the device.

Since this process offers doping profiles and material properties unobtainable by other methods, it is likely to remain in use for the foreseeable future.

5.6 Ion implantation

As integrated circuits became progressively reduced in size, the process of pattern definition using optical methods ceases to be adequate and resolving features below about two or three microns becomes impossible. In addition, problems arise in the diffusion process due to 'undercutting' the etched windows owing to lateral spreading of the junction. Further problems occur which include etching imperfections and 'bowing' of the silicon slice due to heating. Because of these difficulties it is preferable to fabricate devices at lower temperatures and at the same time to improve definition.

Various methods have been used to overcome the above physical limitations, including an extension of the standard photo-etching technique, but using shorter wavelengths (ultra-violet) in order to increase resolving power. Other techniques used are X-ray scanning and ion implantation. A brief description of the latter method now follows.

Ion implantation is an alternative method of junction formation to diffusion, involving the injection or implanting of charged ions into a solid (usually a semiconductor). This method may be used to produce either N- or P-type junctions.

Work in this field was first carried out at Bell Laboratories by R.S. Ohl and W. Shockley during the late 1940s and early 1950s, and in 1954 Shockley patented the use of ion beams for producing the buried base layer in a bipolar transistor.[17]

The application of ion implantation to microelectronics started in 1964–65 when it was realised that buried layers and P–N junctions could be made, the bipolar transistor being one of the first devices to be considered.[18,19] Like thermal diffusion, ion implantation allows accurate control over both the concentration and position of the impurities. It has the following advantages over the standard diffusion process:

(i) A wider range of dopants is available.

(ii) Impurity profiles can be varied over a greater range.

(iii) No sideways penetration of the dopant under the mask edge occurs, as is the case with diffusion.

(iv) Junction profiles can be controlled independently of temperature.

(v) Uniformity of impurity concentration over the wafer and good reproducibility from wafer to wafer and batch to batch.

(vi) Relatively low processing temperatures are feasible reducing the diffusion of unwanted impurities and restricting crystal damage.

The implantation process is followed by annealing to allow the crystal structure to recover, typically at about 650°C for silicon, for a period of about thirty minutes.

In MOS devices, since no sideways penetration of the dopant under the mask edge occurs, precise alignment of the gate electrode with respect to the source and drain is possible and reduces overlap capacitance between drain and gate by an order of magnitude compared with the diffusion process, thus allowing faster switching speeds.[20,21] Also, the ion implantation process makes it possible to fabricate shorter gate lengths, thus reducing the transit time of charge carriers, this again leading to faster switching speeds.

An important feature of the technology of microminiaturisation is that, whichever approach is used, the capital cost of the extremely complex equipment required makes it increasingly difficult for new companies to enter the semiconductor manufacturing field. A further factor of importance is that this technology has led to an increasing erosion of the traditional boundaries between academic disciplines, at various levels. Thus a single individual might now need a knowledge of complicated systems encompassing various aspects of electrical and mechanical engineering, chemistry and physics, including vacuum technology and optics (this being quite apart from a theoretical knowledge of semiconductor device physics.

By the early 1980s ion implantation was being used in every doping step of a typical VLSI (very large-scale integration) process and for a wide range of devices.[22] It is by no means clear however that ion implantation will entirely supersede techniques of thermal diffusion, since, according to Siedel, the shallowest junctions are probably achievable by the use of thermal rather than kinetic energies.[23] Nevertheless, ion implantation, with its stated advantages, will continue to play an important role in device fabrication in future years.

5.7 References

1 PFANN, W.G.: 'Principles of zone refining', *Trans. Amer. Inst. Mining & Metallurgical Eng.*, July 1952, p. 190

2 PFANN, W.G.: 'Zone refining' *Scientific American*, December 1967, p. 62

3 *Ibid.*

4 *Ibid.*

5 BRAUN, E. and MACDONALD, S.: 'Revolution in miniature' (CUP 2nd ed., 1982), p. 43

6 IRVING, S.M.: 'A plasma oxidation process for removing photoresist rilms', *Solid State Technology*, 1971, **14**, p. 47

7 IRVING, S.M., LEMONS, K.E., and BOBOS, G.E.: 'Gas plasma vapour etching process', US Patent No. 3 619 956. Filed 27th March 1969; Patent granted 26th Oct. 1971

8 PENN, T.C.: 'Forecast of VLSI processing—A historical review of the first dry-processed IC' *IEEE Trans.*, 1979, **ED-26**, p. 640

9 FULLER, C.S. and DITZENBURGER, J.A.: 'Diffusion of donor and acceptance elements in silicon', *J. Applied Phys.*, 1956, **27**, p. 544

10 PETRITZ, R.L.: 'Contributions of materials technology', *Proc. IRE*, 1962, **50**, p. 1029

11 LEE, C.A.: 'A high frequency diffused base germanium transistor', *Bell Syst. Tech J.*, 1956, **35**, p. 23–24

12 TANENBAUM, M. and THOMAS, D.E.: 'Diffused emitter and base silicon transistors', *Bell Syst. Tech. J.*, Jan. 1956, pp. 1–22

13 PETRITZ, R.L.: *ibid.*, p. 1029

14 THORNTON, C.G. and ANGELL, J.B.: 'Technology of micro-alloy diffused transistors', *Proc IRE*, 1958, **46**, pp. 1166–1176

15 FROSCH, C.J. and DERICK, L.: 'Surface protection and selective masking during diffusion into silicon' *J. Electrochemical Soc.*, 1957, **104**, p. 547

16 LOOR, H.H., CHRISTENSEN, H., KLEIMOCK, J.J. and THEURER, H.C.: 'New advances in diffused devices'. Presented at the IRE/AIEC Solid State Device Research Corp. Pittsburgh Pa., June 1960

17 SHOCKLEY, W.: 'Forming semiconductor devices by ion bombardment'. US Pat. No. 2 787 564, 28th October 1954

18 KELLETT, C.M., KING, W.J. and LEITH, F.A.: 'High energy implantation of materials'. Scientific Report No. 1, Ion Physics Corporation, AP 635 267, May 1966

19 LEITH, F.A., KING, W.J. and McNALLY, P.: 'High Energy Implantation of Materials'. Final Report, Ion Physics Corporation, AD 651313, Jan. 1967

20 BOWER, R.W., DILL, H.G., AUBUCHAN, K.G. and THOMPSON, S.A.: 'M15 field effect transistors formed by gate masked ion implantations', *IEEE Trans.*, 1968, **ED-15**, pp. 757–761

21 STEPHEN, J.: 'Ion implantation in semiconductor device technology', *Radio & Electron. Eng.*, 1972, **42**, (6)

22 SEIDEL, T.E.: 'Ion implantation', *'VLSI technology'* (McGraw-Hill) p. 260

23 *Ibid.*

Chapter 6

Review of major factors affecting the growth of the United States semiconductor industry

This chapter first carries out a general review of the major factors affecting the growth of the United States semiconductor industry, and then proceeds to consider its more specific aspects. This has been done by dividing the material into a number of sections, each reviewing a particular characteristic of the industry's development.

One important aspect, exhibiting strong distinguishing features when compared with parallel developments in Europe and Japan, is the way in which individual manufacturing companies have developed within the industry. This is dealt with in the Section 6.2. Further sections in this chapter deal with structural, technical, social and geographical aspects of the industry. The object of this approach is to highlight factors peculiar to the US semiconductor industry in order to facilitate a comparison with the parallel growth of the semiconductor industry taking place in other countries. Finally, in view of the outstanding success of Texas Instruments Inc as the major producer of semiconductor devices over three decades, a brief historical survey of this company is included, in order to highlight possible factors contributing to its success.

6.1 General review

From its early beginnings in the 1950s, the semiconductor manufacturing industry has rapidly become one of the major industries of the world, and the semiconductor industry of the United States has, during this time, dominated it. However, growth of the industry within the United States should be viewed within the context of the rapid development of the American electronics industry as a whole. Fig. 6.1 charts the growth of the total sales in the US electronic industry over the period 1947–74. It can be seen that the general trend is an upward curve, indicating an overall steady expansion following the end of the Second World War. Closer inspection indicates, however, that this growth has followed a somewhat uneven pattern, for instance the period 1969–71 exhibited a marked decline, whilst the following years showed a definite recovery. These divergences from the pattern of steady upward growth tend to coincide with the level of government investment in the industry, this in turn depending largely upon political factors, both internal and external to the United States.

Government expenditure in semiconductor research and manufacture has certainly led to political advantages in the arms race. The importance of the military market in stimulating rapid technical change has been perhaps the most significant factor in bringing about the creation of the semiconductor industry within the United States, providing it with an assured market during its vital early stages, in addition to extensively funding research, development and, most importantly, production technology. It has been instrumental in shifting the emphasis from thermionic valve manufacture to solid-state device fabrication, and its requirements have to a great extent dictated the rate at which this replacement has occurred.

Although the military market formed an appreciable proportion of total semiconductor sales during the 1950s and 1960s (see Fig. 6.2, charting the percentage of US semiconductor production designated for military use), sales in the commercial and industrial fields since then have proportionally increased, largely due to the development of new markets and a vast expansion of existing ones, in particular computing. However, even in this field the original stimulus came not from industry but from the needs of the armed forces and in particular the US Air Force.

The 'pump-priming' action of supplying a market for devices at the stage when high unit costs would have precluded their sale in the commercial or industrial market was perhaps the most crucial factor in the development of the industry, and it is significant that this advantage has not been available to semiconductor manufacturers in other countries, which have not only been denied the lucrative American defence contracts, but have received comparatively little support from their own governments. Operation of the 'pump-

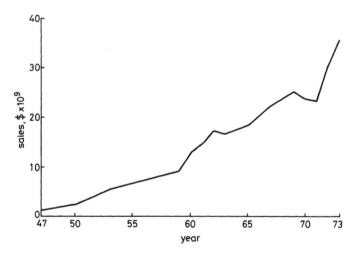

Fig. 6.1 Total sales of US electronics industry (Source: *Electronics* Yearly Market Data Survey; and Electronic Industries Yearbook, 1964)

priming' process with respect to the US integrated circuit manufacture is illustrated in Figs. 6.3 and 6.4, which show clearly the extent to which military contracts absorbed the initial high unit cost of these devices.

One way in which the US semiconductor industry differs from that of all other countries producing semiconductors is that it arose and flourished in conditions which offered little external competition except in specific areas, i.e. cheap germanium devices for transistor radios, and which provided a large and growing internal market for its products. As a consequence of this, the early semiconductor industry concentrated upon supplying the home market, rather than exporting. Nevertheless, problems of overproduction quickly arose, the result of this being that, in a fast developing industry with rapidly falling prices, manufacturers were often left with large stocks of obsolete unsalable devices which then had to be written down in value. For instance, as early as 1954 the industry had sold one million transistors and by 1955 had the capacity to produce 15 million annually, although only 3·6 million were actually made, suggesting that the dangers of this situation arising were quickly realised.[1]

A particular feature of the American semiconductor industry has been the intense competition between rival companies, resulting in rapidly falling prices once volume production of a particular product has been established. Once this situation develops, profit margins are correspondingly slim. This has resulted in continued efforts to develop new products and also to improve the characteristics of existing ones. Not surprisingly, there is a strong emphasis on research, development and production technology within the industry and these areas have received considerable funding. Another factor has been that as the industry has grown in size, it has been able to sustain a number of firms whose function is to service production facilities by supplying specialised equipment of an advanced and often highly expensive nature, such as diffusion furnaces, circuit printing and ion implantation equipment. As production techniques

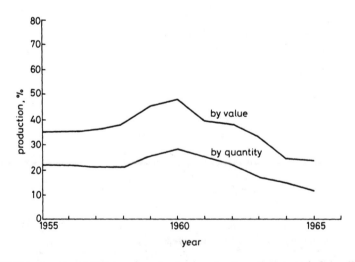

Fig. 6.2 Percentage of US semiconductor production designated for military use (Source: KRAUS, J.: 'An economic study of the US semiconductor industry p. 88)

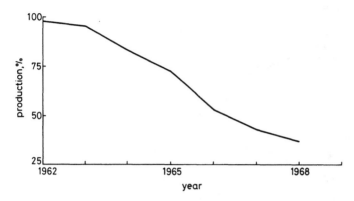

Fig. 6.3 Percentage of US semiconductor production of integrated circuits designated for military use (Source: Reference 12)

become increasingly sophisticated, these firms have assumed relatively greater importance.

Early semiconductor manufacture in the United States was initiated by the thermionic valve companies, although new companies rapidly entered the scene dedicated exclusively to the manufacture of solid state devices. Two factors of importance in assisting this development were an adequate supply of risk capital and the availability of sufficient numbers of highly trained and geographically mobile physicists, metallurgists and engineers, anxious and ready to work in this new field. It is arguable that the emergence of a semiconductor industry stimulated the thermionic valve industry to renewed

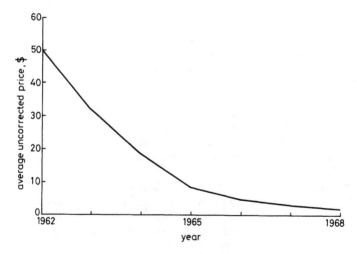

Fig. 6.4 Average uncorrected price of integrated circuits (Source: EIA Market Databook, 1973, Washington, DC)

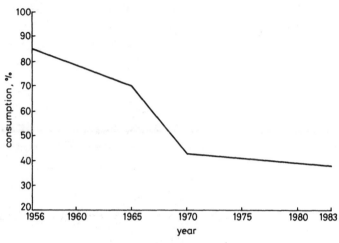

Fig. 6.5 US internal domestic consumption of semiconductors as a percentage of total world consumption (Source: Dataquest, May 1959; and Reference 2)

efforts, particularly in the areas where it has been difficult or impossible for solid state devices to replace their thermionic valve equivalent, for example, in the microwave field and in the transmission of very high powers, thus delaying, where possible, the introduction of solid state replacements.

Although the products of the early semiconductor industry could be considered merely as replacements for a limited sector of the thermionic valve market, this was soon no longer the case. Transistors and other semiconductor diodes have made possible applications which were originally quite unforseen, such as hand-held calculators and digital watches. The semiconductor has therefore, in addition to replacing the thermionic valve in existing markets, created entirely new markets of its own.

From about 1960 onwards, the American semiconductor industry began to supply the then rapidly expanding overseas market for semiconductors. This strategy involved setting up offshore assembly plants in countries where wage rates were lower than in the United States. The American semiconductor industry thus began to operate on an international scale. In 1960, only five offshore plants were in existence. However, a decade later the number had risen to 29.[2] In this activity they have been most successful, and (with the notable exception of Japan) they have progressively increased their share of the world semiconductor market, both by direct exports form the United States and also by their offshore assembly activities.

A notable aspect of the semiconductor industry in America has been the relative decline of internal domestic consumption as a percentage of world consumption (see Fig. 6.5). Although production by US based companies as a percentage of world consumption has also gone into a relative decline since the inception of the industry, this trend has only occurred at a much later stage, and in a more abrupt manner (see Fig. 6.6), largely as a result of Japanese competition. It can be seen from these two Figures that the relative decline in internal domestic production as a percentage of world production has been significantly less than the decline in domestic consumption. This fact illustrates the occurrence of a highly important shift in the nature of the industry from one which in its early years produced devices predominantly for domestic consumption to an industry currently operating on an international scale, fabricating, assembling and selling a significant proportion of its products overseas. In order to compete internationally in this labour-intensive industry, an obvious advantage was (and still is) to undertake the assembly process (where labour costs are highest) in countries where wage rates are much lower than in the United States, whilst carrying out the more complex process of fabrication in the parent country. This strategy is assisted by the relatively low cost of transportation involved.

From 1963 onwards, in addition to the existing fabrication plants already by then situated in Europe, assembly facilities were rapidly set up, mainly in countries where wage rates were low compared with those in the United States. By 1974, 56 of these were in operation in the Far East, 24 in Latin America, and 21 in Europe.[3] (Although by this time six American fabrication plants were operating in Japan, it is noteworthy that no production facilities entirely devoted to the assembly of semiconductors had by then been established in that country).[4]

The decade 1964–74 thus witnessed a significant change in the character of the rapidly growing American semiconductor industry, which now looked increasingly overseas in an effort to expand its activities.

An advantage possessed by the US semiconductor industry has been that at least until very recently they have succeeded in retaining their initial technical lead in process technology. This is particularly important in view of the fact that, owing to the high volume nature of the production process, cost per unit can fall extremely rapidly once a given device is in production, enabling the market to be flooded with these devices at an extremely low cost compared with that only a short time before. Any competitor wishing to enter this market at a later date would consequently find it an extremely costly exercise. (As an

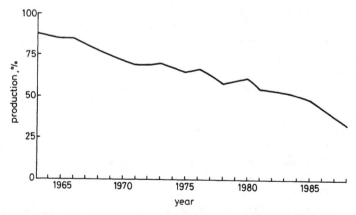

Fig. 6.6 Production by US based companies as a percentage of world consumption (Source: Reference 3; and Dataquest, 1980, 1985–88)

illustration of **this,** a Japanese manufactured dynamic read-only memory (256 Kbits) **retailed** at $110 in 1983, $10 in 1984 and $3 in 1985.)[5] Furthermore, the maintenance of this lead is considered a matter of national importance. In this context, the *Electronics* magazine's 'market trends' spokesman in January 1962 wrote:

> 'In components, the US will always maintain a level (over Europe) due to advanced R & D work in areas funded by the Department of Defense.'

A consequence of this technological lead has been that American semiconductor firms have obtained an additional source of income through licensing arrangements with overseas competitors. Braun and Macdonald state that the Japanese semiconductor industry in the 1970s paid roughly 10% of semiconductor sales in royalties to American companies.[6] In general, American firms have been reluctant to supply the corresponding production technology 'know-how' to back up patent issues, this knowledge being vital in getting a particular device into production with the minimum of delay. Nevertheless, in view of the growth of the Japanese and European semiconductor industries, liberal attitudes towards licensing have been challenged. For instance, as early as January 1963, an edition of *Electronics* 'market trends' states that:

> 'Experts advise the granting of licences only where a market will not be lost as a result. Some firms will be less anxious to grant them in future.'

The author continues:

> 'Executives with international experience point out that foreign firms tend to gather generous amounts of technical information from American firms largely by sophisticated probing techniques used in visits to our plants and talks with our engineers. It is believed that we should try to gain more technical information from European firms than has been our habit, using of course, ethical methods.'

Certainly the actual value of technical manufacturing knowhow is of prime importance; on the other hand the value of patents in the semiconductor industry has been questioned on the grounds that, in such a rapidly changing industry, by the time a patent is granted, the technological innovation is often already obsolete.[7]

In the early days of transistor production, specifically up to about the mid-1960s, the capital cost of setting up a production line and producing relatively unsophisticated devices was low enough to encourage firms to enter the industry. Because of the rapid development of technology, investment in automated plant was low and assembly lines were organised on a labour intensive basis. With the development of the planar process, involving batch processing, the maximum cost was that of final assembly following scribing and breaking of the silicon slice. Labour costs from this stage of the process onwards became all-important, and overseas shipment for final assembly therefore became highly desirable. More recently, with the development of the integrated circuit, capital cost of assembly equipment, e.g. ion implantation apparatus, clean rooms and automatic testing equipment, has greatly increased. The possibility of new firms entering the industry has correspondingly diminished.

When considering the development of the US semiconductor industry, unlike that of Japan or Europe (where the original thermionic valve companies have survived to become the major transistor manufacturers), it has followed a pattern in which the original thermionic valve manufacturing companies quickly lost their pre-eminence to firms specifically specialising in semiconductor fabrication. However these firms, once established, have since largely remained in the forefront, although with some variation in their respective fortunes. Because of the aforementioned technical developments resulting in high capital cost of assembly plant, it is now becoming increasingly unlikely that significant changes in the structure of the semiconductor industry (apart from takeovers and mergers) will take place in the foreseeable future, the possible exception being the formation (in Europe) of government assisted ventures along the lines of Inmos.

A decided advantage possessed by the American semiconductor industry over its rivals has been the so-called 'brain drain'. Since the inception of the industry, scientists and technologists of ability have been attracted to the United States, bringing with them not only their expertise, but also information on research, development and production techniques from their country of origin. This development has tended to be a one-way process, and has certainly not worked to the advantage of the European semiconductor companies.

Also, when considering the growth of the semiconductor device industry, it is important to realise that its fortunes have been closely connected with that of computers, particularly since the development of the integrated circuit. The rapidly growing computer industry has been highly dependent upon semiconductor integrated circuits since their development in the early 1960s, and the integrated circuits themselves have been in turn largely dependent upon the continued development of computers, a situation described by Mackintosh as 'Industrial Synergism'.[8] He argues that the success of the American semiconductor industry is largely due to this fact and, writing in 1978, cites the relative lack of success of the Japanese and European semiconductor industries at that time as being due to the absence of synergistic users of semiconductor components.[9]

Finally, perhaps the most significant development to occur since the establishment of the American semiconductor industry has been the emergence of successful Japanese competition during the present decade, particularly in the field of very large scale integration (VLSI). This development has exposed a relative weakness in the American economic structure, namely, the inability of the industry to match the Japanese success in raising long term low interest loans. These loans have enabled the Japanese semiconductor industry to sustain substantial financial losses over periods of several years, whilst continuing to mass-produce devices at low cost. Furthermore, this situation has been compounded by the considerable technical lead now held by Japan in some areas of integrated circuit manufacture.

The only possible remedy for the American semiconductor industry, faced with this new situation, appears to be substantial Government aid, together with the imposition of some form of import control. A report published by the

Defense Semiconductor Dependency task force of the United States Defense Science Board has recommended that 'the U.S. Government invest $2·2 billion to ensure the USA is never dependent on foreign suppliers for DRAM and other memory chips'.[10] The threat of import control by the United States resulted in the signing of an agreement with the Japanese Government in July 1986 regarding trade in semiconductors, to run for five years, the intention of the agreement being 'to prevent Japan from dumping ICs and undercutting prices in overseas markets, while requiring Japan to buy more imported semiconductors'.[11]

6.2 Developmental aspects—patterns of industrial growth

Following the invention of the point-contact transistor at Bell Laboratories in December 1947, patent applications were filed in February 1948. A military preview was given in June, followed a week later by the public announcement of the invention on 30th June of that year. Manufacture of point-contact transistors began on a production line basis at Western Electric around October 1951. In April 1952 Bell Laboratories, adopting an open-door policy from the start, held an eight-day technology symposium to explain how point-contact transistors could be made,and also to reveal information regarding work in progress on junction transistors; the information made available at his time included detail and processing information involving methods of crystal growing and device construction. This first Bell symposium was attended by 35 firms, including ten from overseas, each paying an entrance fee of $25 000, the amount to be deducted from the licensing fees.

The influence of Bell Laboratories, particularly at this early stage, can hardly be too strongly emphasised. This influence consisted mainly in the dissemination of technical know-how, backed by a liberal licensing policy. Following this, Bell arranged a second important symposium which was held in January 1956, and in which information on oxide masking and diffusion technology was communicated to 72 industrial firms who were by now Bell licensees.

The first firms to become involved in semiconductor manufacture in the United States, following the lead of Western Electric, were all thermionic valve manufacturing companies, namely the Radio Corporation of America (RCA), Sylvania, General Electric and Raytheon, all of which were engaged in semiconductor manufacture by 1952. By April of that year, the junction transistor was being manufactured at Western Electric in limited quantities, and also during the same year General Electric manufactured the first germanium alloy devices, this invention being the first major technical innovation to take place outside Bell Laboratories.

However, the thermionic valve companies were quickly challenged by a number of new semiconductor manufacturers who engaged solely in this activity, the most successful of all being Texas Instruments Inc. of Dallas, Texas (which in May 1954 announced the successful fabrication of the grown junction silicon transistor, and which was to become from 1957 to 1985 the world's largest manufacturer of semiconductor devices). These new companies, with their more aggressive marketing policies, quickly seized a substantial share of the American transistor market, progressively increasing their influence over the next decade, as can be seen from Table 6.1. Mackintosh, writing in 1978, states:

> 'only one of the top ten US vacuum tube manufacturers in 1955 (RCA) has survived as a significant IC producer today.'

There is also evidence that structural reasons within the valve companies contributed to their failure to maintain their leading position as semiconductor manufacturers. For example, Golding writes that:

> 'In all but two cases (GE and Sylvania) the corporate management exacerbated the conflict between vested interest and the new processing method by placing the embryo semiconductor activity within the administrative confines of the valve division. This had the effect of allocating responsibility to those with an intellectual (and emotional) investment in a totally different current technology.'[12]

Another inhibiting factor was that, in the case of valve manufacturing companies with system assembly requirements (such as RCA, GE and Westinghouse), a conflict of interest lay between satisfying these internal requirements and those of the open market.

What is apparent from Table 6.1 is the rapid change in fortune overtaking the major companies (with the notable exception of Texas Instruments Inc.) engaged in semiconductor device production, the principal factors accounting for this fluctuation appearing to be the unstable situation arising where high volume production of a low cost product undergoing rapid technical change places stringent demands upon development, production and marketing of the product. A weakness in any of these areas will be rapidly exploited by competitors.

In spite of the varying fortunes of the leading semiconductor companies, the actual number of firms present in the industry has remained remarkably constant over a long period. By 1957 there were approximately 50 companies engaged in semiconductor manufacture, including the fabrication of solid-state diodes and rectifiers. This number quickly built up to a plateau in 1962, and has remained at approximately the same level ever since, in spite of continual entry and departure. For example, Braun and Macdonald mention that even though a number of companies had entered and left the industry, by 1972 there were about 120 semiconductor companies manufacturing transistors, diodes and rectifiers, approximately the same number as a decade before. Figures published by Dataquest in 1985 indicate that about 115 American owned semiconductor companies were operating in the United States in that year. Despite the fact that there existed such a large number of companies, the market share held by the top ten of these organisations has consistently been far in excess of that held by the remainder (many of whom concentrate on supplying a specific niche within the market). This situation is illustrated in Table 6.2.

With the development of the integrated circuit, progress techniques rapidly became more sophisticated, and the capital cost of setting up a semiconductor manufacturing facility became correspondingly greater. Over at least the last

Table 6.1 Leading US discrete semiconductor manufacturers in order of ranking

1955 (tubes)	Ranking	1955	1957	1960	1963	1965	1966
RCA	1	Hughes	TI	TI	TI	TI	TI
Sylvania	2	Transitron	Transitron	Transitron	Motorola	Motorola	Fairchild
GF	3	Philco	Hughes	Philco	Fairchild	Fairchild	Motorola
Raytheon	4	Sylvania	GE	GE	GE	GI	WE
Westinghouse	5	TI	RCA	RCA	WE	WE	GE
Amperes	6	GE	WE	WE	RCA	GE	GE
National Video	7	RCA	Motorola	Motorola	RCA	RCA	RCA
Ranland	8	Westinghouse	Raytheon	Clevite	Westinghouse	Sprague	Westinghouse
Eimac	9	Motorola	Sylvania	Fairchild	Philco	Philco	GI
Lansdale Tube	10	Clevite	Westinghouse	Sylvania	TRW	Transitron	GM

GE: General Electric
GI: General Instruments
GM: General Motors
RCA: Radio Corporation of America
TI: Texas Instruments
TRW: Thompson Ramo & Wooldridge
WE: Western Electric (in-house sales only)

Source: TILTON, J.: 'The international diffusion of technology: a case study of the semiconductor industry' (The Brookings Institution)
MACKINTOSH, I.M.: 'Large scale integration, intercontinental aspects', *IEEE Spectrum*, (June 1978), **15**, p. 54

decade, this economic trend has inhibited entry into the industry and introduced a measure of stability previously lacking. Fluctuations in the fortunes of the major semiconductor manufacturing companies have also tended to stabilise, as may be seen from Table 6.3. The exception is Fairchild, taken over by Schlumberger (France) in 1979.

6.3 Structural aspects—role of spin-off companies in the development of the industry

A feature of the American semiconductor industry markedly distinguishing it from that of Europe and Japan has been the existence of a relatively large number of spin-off companies from existing semiconductor manufacturing firms. A spin-off company may be defined as one which has obtained its technology as a result of the direct transfer of personnel from a parent company. Golding, writing in 1971, estimated that spin-offs outnumbered new ventures on the part of existing semiconductor concerns by a ratio of about four to one.[13] This phenomenon appears to have been a consequence of a high level of personal mobility coupled with the ready availability of an adequate supply of risk capital. A further factor assisting this trend and which appeared to have been significant was, according to Golding, the availability of Federal Government research and development procurement contracts. However, he quotes a US Department of Commerce report to the effect that by this time (1971):

> 'it is apparently more difficult for smaller competitors to obtain military/space contracts despite the existence of a 'set aside' procedure which purports to set on one side contracting opportunities for small businesses'.[14]

Certainly since that date, the formation of new spin-off companies has significantly decreased. The two major factors contributing to this trend appear to have been a general reduction in the flow of available risk capital and also the

Table 6.2 Percentage market share held by the ten leading companies in the USA

Year	1957	1960	1963	1966
Market share (%) (top ten companies)	77	76	72	82
Share of market (%) held by top company (Texas Instruments)	20	20	18	17

Source: TILTON, J.: 'The international diffusion of Technology: a case study of the semiconductor industry' (The Brookings Institute)

Table 6.3 Leading US discrete and IC semiconductor manufacturers in order of ranking

Ranking	1975	1978	1981	1984	1985	1986	1988
1	TI	TI	TI	TI	TI	Motorola	Motorola
2	Fair child	Motorola	Motorola	Motorola	Motorola	TI	TI
3	Nat. Semi.	Nat. Semi.	Nat. Semi.	Nat. Semi.	Intel.	Nat. Semi.	Intel.
4	Intel.	Fair child	Intel.	Intel.	Nat. Semi.	Intel.	Nat./Fairchild

rapidly increasing capital cost in setting up a contemporary semiconductor manufacturing facility. In general, spin-off companies have tended to be short lived. Braun and Macdonald, quoting a 'Dataquest' survey, state that:

> 'of thirty-five semiconductor companies started between 1966 and 1975, only seven remained independent in 1980, and no fewer than seventeen merchant semiconductor companies were then owned by conglomerates with primary interests outside semiconductors.'[15]

This quotation highlights the importance of the role of the small spin-off companies in making available the necessary expertise to enable larger organisations outside the industry to enter it. In particular, this route of entry provides a convenient path for organisations with specific interests in the industry, such as systems manufacturers. Alternatively, the acquisition of small firms engaged in advanced state-of-the-art fabrication by older companies as an alternative to in-house research and development, or in an attempt to catch up in the latest technology, has been attempted, an example of this being the purchase in 1966 of General Microelectronics by Philco, which was used to reinforce the latter company's position in the MOS market.

Golding states that the great majority of small companies which entered the American semiconductor industry originated as spin-off firms, and traces the development in detail back to Bell Telephone Laboratories and a number of additional primary sources. Certainly the original or primary spin-off companies from Bell were more successful than the existing thermionic valve companies in establishing themselves as high-volume semiconductor device manufacturers. They include Texas Instruments (1953), IBM (1962), Sylvania (1953) and Transitron (1952). Secondary spin-off companies of importance arising from primary spin-offs include Fairchild (1957), a spin-off from Shockley Transistor, which itself was a primary spin-off from Bell in 1955. Texas Instruments generated in its turn Siliconex (1962) and Mostek (1969), whilst Fairchild produced more spin-off companies than any other, these including Signetics (1961), General Microelectronics (1963) (which later became Philco–Ford in 1966) and National Semiconductor in 1967. Braun and Macdonald mention that 'by the early seventies, some forty-one companies had been formed by Fairchild employees'. An exception was the Motorola Company, formed as a spin-off from General Electric in 1957.

6.4 Technical and social aspects

6.4.1 Position of research and development
Although the semiconductor was a product of fundamental science, its subsequent history has largely been that of successful technical innovations. As a consequence, emphasis rapidly shifted from basic research and development to the area of production development. Golding estimated that by 1965 an appreciable decline in real terms had taken place in investment in the field of research and developments since 1959.[16] Braun and Macdonald state that research and development expenditure as a proportion of sales had dropped from 27% in 1958 to perhaps as low as 6% in 1965[17] and by 1972:

Table 6.4 Fundamental advances originating at Bell Laboratories

Specific innovation	Date
Single crystal growth	1949–53
Zone refining process	1950–51
Ge point-contact transistor	1951
Si alloy junction transistor	1952
Oxide masking	1957
Diffusion	1950–57
Silicon mesa transistor	1958
Epitaxial transistor	1960

'of about thirty six million dollars spent by the semiconductor industry on basic research and development, some twenty-eight million dollars, or 80%, was spent by Bell Laboratories and IBM, neither of which sold in the open market.'[18]

It would appear that several factors were responsible for this state of affairs, one of the more important being the development of rapid product cycles in an extremely competitive industry, where profit margins on a particular device would tend to fall to a low level fairly quickly. This occurrence resulted in a considerable emphasis upon getting new devices into production as soon as possible and then improving the product whilst already in production on an *ad-hoc* basis. The shift from military to non-military products during this time, with the emphasis on high volume production of cheap germanium devices could hardly have failed to accentuate this trend. Thus, during the 1950s and 1960s, the emphasis shifted from basic research towards applied research and development and from the laboratory into the production area. During this period, only large organisations, such as AT & T, could afford to sustain an appreciable measure of basic research and development effort. However, engaged as they were in the specific area of telecommunications and producing devices through their subsidising Western Electric purely on an in-house basis, AT & T had priorities of a somewhat different nature from those of other semiconductor manufacturers.

Consequently, it would seem that the rapid advances made in semiconductor technology during the 1950s and 1960s were not so much a result of traditional Research and Development activity, but primarily due to ongoing process development. Nevertheless, a substantial number of fundamental innovations upon which process development ultimately depended were due to specific research and development effort, and in this area the role of Bell Laboratories was considerable. It is notable that a large number of fundamental advances in device technology owed their origin to that organisation. This is illustrated by Table 6.4 which gives some indication of their early achievements.

Although, as stated above, the major emphasis of product development in the semiconductor industry has lain in its production area, rather than in separate research and development, it is important to note that nevertheless considerable funding by the American Government has also taken place in the latter area. Hogan estimates that Government support for research and development carried out between 1958 and 1977 amounted to 900 million dollars[19] and there can be little doubt that this funding has considerably assisted the semiconductor industry to maintain its technical advantage and, in the words of Mackintosh; 'grossly disturb normal competitive conditions and commercial criteria in this industry'.[20] Writing at an earlier date (1973), Mackintosh states that:

> 'Clearly the most important factor in providing the US technical lead was the tremendous research and development funding provided through the 1960s by the US government.'[21]

but nevertheless continues in the same article:

> 'It is important to note that the bulk of these funds was channelled to industry, not the universities or the government research laboratories.'

The nature of industrial growth and development of the semiconductor industry has been that it has consistently placed a major emphasis upon getting a technically sophisticated product rapidly into high volume production. It is in this area in particular that American engineering has consistently outperformed that of Europe. Certainly this is widely recognised and is a key factor contributing to the success of the American semiconductor industry. It is therefore instructive to compare the situation with Europe in this respect and to seek reasons why this difference in performance should exist. One factor noted by observers is that significant social differences exist between European and American Society and that this situation is reflected within the semiconductor industry. For instance, Braun and Macdonald, reporting a conversation with W. Corak, write:

> 'In the United States, those engaged in industry are held in as much esteem as those working in universities or government research laboratories. In Europe this is definitely not the case.'

Having worked for both British and American semiconductor companies, the author would certainly concur. Conversations suggest that in Japan, like the United States, there also appears to be parity of esteem between research and production.[22] Parity of esteem between R & D and production facilitates exchange of personnel between both areas to the benefit of each, and inhibits the 'brain drain' from production to R & D which might otherwise occur. Other important factors in favour of the American semiconductor industry are the sheer size of the operation and also the number of trained engineers and scientists upon which the industry can draw. For example, it was estimated in 1966 that there were approximately 10 000 semiconductor engineers in the United States and about 300 in Britain[23] (and, most importantly, not only more production engineers but consequently a much larger number of first-rate production engineers). Also, in a mobile society such as America, it is most likely that on average the better engineers will gravitate to the more successful companies, further improving the performance of those companies. A more mobile workforce than that existing in Europe enables technical knowledge to be rapidly diffused both from research and development to production and vice versa and between one semiconductor company and another. In this context Mackintosh writes:

> 'The United States has a major advantage in the fortuitous combination of a high rate of personnel mobility with the existence of several large and highly capable research laboratories which have acted as national generators of technology and technologists. Thus, in America, the diffusion of technology has occurred mainly through the diffusion of people.'[24]

This diffusion of knowledge has been considerably assisted by the presence of skill clusters. In view of the importance of this development, what follows next is a review of the factors affecting the location of the semiconductor industry, including the establishment of skill clusters and also the significance of overseas development.

6.4.2 Role of silicon in the development of the industry

A feature of the American semiconductor industry is that it has grown up within a high wage economy, equipment assembly costs being relatively high when compared with the situation in Europe and to an even greater extent in comparison with that in Japan. Consequently, in the period before offshore operations were undertaken, the Japanese and, to some extent, European manufacturers were able to enter successfully the American cheap portable radio market. Japanese manufacturers, in particular, made large inroads into this area at an early date. For example, in 1955 shipments of radio, television and associated parts to the United States totalled $230,000. By 1959 this figure had risen to $55 million. Between 1958 and 1959, Japan achieved a 244% increase in dollar value shipments to the United States of radio and television apparatus and parts, and by the latter year had captured 50% of the American market for transistor portable radios.[25]

Engaged in producing large quantities of cheap germanium devices during the 1950s and 1960s, Japanese and European semiconductor firms tended to be relatively strong in the field of consumer electronics, which was largely based on a germanium technology. Defence industry requirements in these countries were not sufficiently strong to exert a significant market pull towards an alternative silicon technology.

The situation in America was significantly different, however, the prevailing condition during this period being that of relatively high labour costs, together with extensive military requirements. These military requirements were of such a nature that they could be best satisfied by adopting a silicon based technology. As a consequence of this the American semiconductor industry rapidly dominated in this field and, in addition, silicon devices became essential to the construction of military and computer equipment. This situation is an example of what is described by Mackintosh as 'Industrial Synergism', which he defines as:

'the mutual interdependence of different industrial sectors and, in particular, of the equipment and component sections of the electronics industry.'

This state of affairs was further reinforced by the invention of the silicon planar transistor, rendering all previous process technologies obsolete. Since the planar process cannot be adapted to a germanium based technology, the Japanese and European semiconductor industries were forced to follow suit during the 1960s, to abandon germanium and to adapt to the manufacture of silicon planar transistors. This changeover was not without its difficulties, the consequence of this being to render a further advantage to the dominant American industry. Another important spin-off from the American predominance in silicon technology was that it placed the US semiconductor companies in an advantageous position with regard to integrated circuit development, thereby assisting the maintenance of the existing technological lead over foreign rivals.

6.5 Geographical aspects—skill clusters within the US semiconductor industry

The growth of skill clusters has been a key factor in the development of the US semiconductor industry, and this highly significant event contrasts strongly with the situation in Europe, where no comparable development has yet taken place.

The initial location of the US semiconductor industry in the early 1950s was determined, as in Europe, largely by that of the existing thermionic valve companies. The original centre of gravity was on the East Coast, in the vicinity of Boston and New York; i.e. close to large conurbations where venture capital, technical services and expertise were more available than elsewhere. Early examples of leading semiconductor firms being set up on the East Coast during this period are Transitron (Massachusetts), Hughes (Rhode Island) and Germanium Products (New Jersey).

However, a change in this state of affairs soon occurred and was most likely due to an attempt to solve an important technical problem common to all transistors before the advent of surface passivation, namely, the effect of surface contamination upon electrical device parameters, due to the presence of unwanted impurities during manufacture. A possible solution to this problem was to fabricate devices in conditions of extra cleanliness, and this idea may well have been significant in view of the decision of some semiconductor companies to move to areas such as Colorado and California and around which skill clusters tended to develop during the next decade. Furthermore, location in areas of this nature would tend to be an additional attraction to prospective employees, and this has been readily recognised by employers, whose advertisements for personnel have usually drawn attention to the advantages of the particular area in question.

Subsequent growth of the semiconductor industry has tended to take place around a few skill clusters rather than spread in a more even fashion throughout the country. Factors responsible for this development include the need for close contact between semiconductor firms and suppliers of materials such as highly pure metals, gases, and manufacturers of equipment specific to the industry. Semiconductor firms working in the same geographical area would now have access to these facilities, in addition to the advantages of clean air and water. As a skill cluster developed in size, other specialised technical and financial services necessary to maintain and support the industry would tend to be attracted to the area, increasing in turn the desirability of the given location. Certainly these factors demonstrate the continuing importance of skill clusters and, following their establishment, indicate the increasing difficulty in setting up a semiconductor manufacturing facility outside these areas. It is significant that, during the last two decades at least, no major semiconductor manufacturing facility has emerged outside the existing skill clusters in the United States. Certainly only a few firms in that country have survived outside large skill clusters and have themselves generally been large and well established, such as Texas Instruments in Dallas, Texas, and Motorola in Phoenix, Arizona.

Further secondary advantages that have arisen as skill clusters have grown in size is that they have facilitated a frequent interchange of staff between individual companies, and this has been a highly important means of diffusing technology. Also important is that the growth of skill clusters has tended to generate a ready supply of high-grade technicians and other support staff. One result of this situation is that the emergence of spin-off companies in skill cluster areas has tended to be frequent. It is also a matter of note that both Texas Instruments and Motorola, geographically well isolated from other semiconductor companies, have not produced any spin-off companies of importance. A further attraction of restricting the concentration of the semiconductor industry to a few major centres is that employees so desiring may indulge in 'job-hopping' without the disadvantages of moving area.

A significant connection between the location of skill clusters and the availability of risk capital appears to exist. The existence of an adequate supply of risk capital has undoubtedly been an important factor in the emergence of the many spin-off companies which have proliferated in the Silicon Valley area. In this context, Braun and Macdonald write:

> 'It is also of no small relevance to the presence of the semiconductor industry in Silicon Valley that the area was one in which risk capital for finance enterprise was more readily available than in the other areas.'

and they continue:

> 'In San Francisco, financial institutions and private sources of venture capital came to understand the semiconductor industry and their support has been fundamental to its growth.'[26]

During the 1960s the centre of the semiconductor industry shifted from the East Coast of the United States to the West Coast. One reason for this shift was the role played by the Fairchild company in initiating the development of a major skill cluster in the San Francisco area popularly known as 'Silicon Valley'. This company, established in 1957, was itself a spin-off from the Shockley Semiconductor Laboratories which had been set up by W. Shockley at Palo Alto in 1955.

Fairchild, unlike Texas Instruments and Motorola, produced in its turn a further number of spin-off companies, many senior staff leaving in the late 1960s. Braun and Macdonald state that:

'At a conference held in Sunneyvale, California, in 1969, of 400 Semiconductor men present, less than two dozen had never worked for Fairchild.'

and:

'by the early seventies, some forty-five companies had been formed from former Fairchild employees.'[27]

Since the advent of the integrated circuit, the major trend in device technology has been in the direction of microminiaturisation. This has largely involved a process of continual improvement in production engineering technology rather than further advances in the fundamental understanding of solid-state physics. With the increasing use of complex and expensive plant, such as ion implantation equipment, a shift away from labour-intensive assembly has occurred. Also, because of the complexity of the new equipment required, firms specialising in its construction have become relatively more important to the industry. These developments have additionally generated the need for a new type of highly trained and highly specialised technician capable, however, of crossing the boundaries of the traditional disciplines.

These technical changes have brought about in their turn an effect upon the geographical distribution of the semiconductor industry. For instance, in view of the shift away from labour-intensive assembly, the advantages of off-shore assembly in areas of low-wage economy may become of less importance than formerly (although in an industry where profit margins rapidly tend to become minimal, the advantages of low-wage assembly must always remain a major factor). These recent developments have, however, reinforced the need for adequate skill clusters and this situation is likely to remain in the foreseeable future.

6.6 Brief history of Texas Instruments (TI) Inc.

The dominant position in the American semiconductor industry has been held since the mid-1950s by Texas Instruments Inc.[28] In view of its somewhat unusual position in this respect, it is of interest to consider the history of this company in order to identify factors which have led to its undoubted success.

The company was originally founded under the title Geophysical Services Inc. (GSI) in 1930, and carried out geophysical exploration on a world-wide scale. In the Second World War, GSI supplied airborne magnetometers to the government, thereby establishing links with military connections. However, during this period it was still a small company, its business with the military totalling not much more than $1m. during the entire period of hostilities. Business turnover during 1946 amounted to a little over $2m., and total company profits for that year were $123 000 after taxes.[29] The Executive Vice President of the Corporation, P. Haggerty, mentions that company policy at that time had three specific aims: firstly, to manufacture and service geophysical equipment, secondly, to re-enter the military equipment field as a supplier of electronic equipment and thirdly to enter non-military markets to develop existing and new technologies.

By 1949, GSI turnover amounted to $8·5m. and profits for that year were $263 000 after taxes; however by 1951 the centre of gravity was shifting away from geophysical operations and the company changes its name to Texas Instruments (TI), although work in the field of geophysical exploration continued.[30]

The advent of the Korean war, beginning in 1950, led to further military contracts in the electronics field. Entering the thermionic valve manufacturing industry was considered at this time, but rejected on the grounds of lack of sufficiently innovative ideas in this well-established field. At this time, the idea that the newly invented transistor would be of great future significance, coupled with the realisation that the successful development of this device would depend upon the fundamental understanding of materials, led the management of TI to plan the following strategy. Firstly, to obtain a patent licence from Western Electric, secondly, to establish a project engineering group to develop, manufacture and market semiconductor devices and, thirdly, to establish research laboratories which would be given the task of carrying out work in the solid-state field with emphasis upon semiconductor materials and devices.

TI representatives attended the 1952 Bell Transistor symposium. Previous to this event, no one at TI had had any previous experience in the semiconductor field. However, in 1952, G. Teal, an employee of Bell, was engaged as director of research laboratories, which he then had to create from scratch. Events moved rapidly—by June 1952, a crystal puller had been built, and point-contact transistors fabricated. Haggerty states that by the end of 1952 their crystal-pulling equipment was 'probably two or three years ahead of anything available to others in the industry'.[31]

By the latter part of 1953, TI were producing germanium transistors in quantity and in the fourth quarter of that year manufactured 7500 of these devices.

The first silicon grown junction transistors were produced by G. Teal in April 1954, and by the following month small quantities of this device were already in production. The decision to produce a silicon transistor was perhaps the most significant step in the history of TI. The limitations of germanium were realised by the management from the start, and the decision to produce silicon was taken fully in the knowledge that, if successful, lucrative military contracts could be readily obtained. Commenting on the successful result of this achievement, Golding writes:

> 'For three years TI was effectively the sole supplier of silicon transistors at prices yielding profit margins which the chairman of the company later described as "exceptional".'[32]

In June 1954, TI entered into a joint programme with the Regency Division of Industrial Development Engineering Associates to create an all-transistor pocket radio. This was done and the first of these went into production in October 1954, selling at $50 per unit. Production built up rapidly. By January

1955, 1500 units were marketed, and by April of that year, 32 000. Within twelve months, over 100 000 had been sold. This feat, according to S.T. Harris, 'changed the world's acceptance of the transistor and did much to launch TI'.[33] However, in spite of this early success, TI did not enter the consumer business at this time, but concentrated exclusively upon the production of devices until 1971.

The rapid and successful growth of the company was reflected in the rise of its share prices. TI shares, worth $5 in 1952, had by 1959 reached a level of $191. It is significant that the slump in germanium prices during the years 1961–63 which affected the industry generally had much less effect upon TI, because of its high proportion of sales in the silicon market.

More recent technical innovations of fundamental importance contributed by TI include the invention of the integrated circuit by J. Kilby, who took out the first patent for such a device. Kilby left Bell Laboratories to join TI in 1958, and produced the first integrated circuit in the Autumn of that year. This device was made of germanium.

A further important development was the concept of a 'computer on a chip', which, according to Braun and Macdonald, almost certainly originated with Texas when G. Boone and M. Cockburn demonstrated its feasibility and applied for a patent, which was granted in 1971.[34]

The dominant position of TI within the semiconductor industry has been largely achieved by internal growth rather than through mergers with other corporations. Unlike most other large American semiconductor firms, it has generated few 'spin-off' companies and this is most likely because of its consistent financial success, together with its high quality of management. However, because of its reputation at management level, TI trained managers have been somewhat in demand and labour turnover at this level on an individual basis has been fairly significant.

Golding notes that unlike most semiconductor companies, TI tended to adopt a policy of mechanisation and automation rather than attempting to reduce labour overheads by offshore assembly, only entering this field in 1967–68, when plants were set up in Taiwan, Singapore and Curacio.[35] However, overseas fabrication plants were set up at an earlier date in Europe, both at Bedford, UK (1957), and at Nice, France. TI Japan was set up in 1968 as a joint venture with Sony, but was later taken over entirely by TI, who by 1957 were producing the majority of their VLSI chips at that location.

A distinctive feature of Texas Instruments is that product development is not solely a function of the research and development departments, but is also very much an on-going process within the production area. Production engineers are encouraged (indeed are required) to carry out continued experimental work in order to improve the product—when employed by TI as a production engineer, the author spent a considerable portion of his time engaged in this activity. Production engineers are also encouraged to become involved with many aspects of company organisation, particularly the costing system, and short courses are frequently arranged on an in-house basis covering a wide variety of topics. The effect of this is to introduce a dynamism probably absent in many other companies, provide excellent management training, and generally stimulate morale. A further significant feature is the professionalism of the marketing department. Marketing engineers are required to be technically proficient in the field of semiconductor applications and are generally able to visit prospective customers and offer technical advice. Although their function in this respect is that of sales engineer rather than that of salesman. nothing is spared in energy and effort to market the product.

6.7 References

1 'Transistors Growing Up Fast', *Business Week*, 5 Feb 1955, p. 87
2 FINAN, W.: 'The international transfer of semiconductor technology through US based firms'. National Bureau of Economic Research, NY, 1975, p. 94
3 Department of Commerce: 'A Report on the Semiconductor Industry' (US Government Printing Office, Washington DC, 1979, pp. 85–86
4 *Ibid.*
5 *Financial Times* Survey, 6 Dec. 1985
6 BRAUN, E. and MACDONALD, S.: 'Revolution in miniature' p. 153
7 *Ibid.*, p. 132
8 MACKINTOSH, I.M.: 'A prognosis of the impeding intercontinental LSI battle', 'Microelectronics into the 80's' (Macintosh Publications, 1979), p. 65
9 *Ibid.*
10 PARRY, S.: 'Time to get tough with Japan', *Electronics Times*, 19 Feb. 1987, p. 1
11 'Solid State', *IEEE Spectrum*, Jan 1987, p. 45
12 GOLDING, A.M.: 'The semiconductor industry in britain and the United States: A case study in innovation, growth, and diffusion of technology'. D.Phil.Thesis, University of Sussex, 1971, p. 175
13 *Ibid.*, p. 249
14 *Ibid.*, p. 240
15 'Dataquest', 15 June 1980
16 GOLDING, A.M.: *op.cit.*, p. 134
17 BRAUN, E. and MACDONALD, S.: *op.cit.*, p. 141
18 *Ibid.*, p. 163
19 HOGAN, C.L.: 'An Address to semicon/Europe 75, Zurich', 4 November 1975
20 MACKINTOSH, I.M.: 'Prognosis of the impending intercontinental LSI battle', *Microelectronics into the 80's*, (Mackintosh Publications, 1979, p. 63
21 MACKINTOSH, I.M.: 'The future structure of the semiconductor industry', *Radio & Electron Eng.*, 1973, **43**, 1973, p. 150
22 Conversations with K. Kitagawa (Engineer, Toshiba Co.) and S. Umitani (Engineer, Komatsu Ltd.)
23 Conversation with F. Thurmond (Chief Engineer, TI, Bedford, 1966)
24 MACKINTOSH, I.M.: 'A prognosis of the impending intercontinental LSI battle', 'Microelectronics into the 80's', (Mackintosh Publications, 1979, p. 67
25 *Ibid.*, p. 63
26 BRAUN, E. and MACDONALD, S.: *op.cit.*, p. 129
27 *Ibid.*, p. 126
28 See Tables 6.1 and 6.3
29 HAGGERTY, P.E.: 'A successful strategy'. *25th Anniv. Observ. Transistor Radio and Si Transistor*, TI Publication, p. 1
30 *Ibid.*, p. 2
31 *Ibid.*, p. 4
32 GOLDING, A.M.: *op.cit.*, pp. 158–159
33 HARRIS, S.T.: 'Marketing the product'. *25th Anniv. Observance Transistor Radio & Si Transistor*, TI Publication, p. 3
34 BRAUN, E. and MACDONALD, S.: *op.cit.*, p. 113
35 GOLDING, A.M.: *Ibid.* p. 160

Chapter 7
Review of the major factors affecting the growth of the Japanese and South Korean semiconductor industries

This chapter deals principally with the growth and general characteristics of the Japanese semiconductor industry, the more recent and relatively less important South Korean industry being considered in somewhat less detail.

Firstly, the economic structure within which Japanese transistor manufacturing has developed is described, and is then followed by an account of the growth of the industry within an environment dominated by the major American semiconductor manufacturers. This environment has to a large extent determined the response of both Japanese manufacturers and Government, and consequently the nature of the growth of the semiconductor industry in Japan is largely bound up with the development of its larger competitor.

The account of the growth of the transistor industry in Japan begins with a review of the original challenge to the American semiconductor industry in the field of cheap germanium commercial devices, and the later entry into the area of silicon device technology is then considered. It is argued that technical developments which were to lead to the recently successful challenge to United States manufacturers, principally in the field of very large-scale integration have resulted from the long-term policy pursued by the Japanese Ministry of International Trade and Industry (MITI). Backed by substantial funding in the areas of both production and device development, this strategy has proved to be a viable successful alternative to the American approach, which has considerably relied upon military contracts and support.

Finally, the characteristics of the newly developed South Korean semiconductor industry are discussed, this country's achievement being of particular interest owing to its singular development from the status of an offshore assembly facility to that of a fully fledged manufacturing industry, competing commercially on a world-wide scale.

7.1 Japanese semiconductor industry

7.1.1 Economic structure of the Japanese semiconductor industry
The recent spectacular achievements of the Japanese semiconductor industry must be considered within the general framework of that country's economy. The growth of the Japanese economy is certainly impressive, about 9% per year (in real terms) in the 1950s, more than 10% in the 1960s and 13 to 14% at the

latter end of that decade.[1] Growth of the Japanese semiconductor industry has paralleled this trend, and now represents a significant challenge to the dominance of the United States in this field. In order to understand this situation it is important to consider the conditions under which this challenge has emerged.

Unlike the situation in the United States, the number of semiconductor companies is relatively small and currently the largest and most important of these are as follows: Nippon Electric Co., Hitachi Ltd., Toshiba Corp., Matsushita Inc., Mitsubishi Elect. Corp., Fujitsu Ltd. and Tokyo Sanyo Inc. All these firms are vertically integrated, manufacturing a wide variety of electronic equipment in addition to other commodities. Again, unlike the situation in the United States, companies manufacturing thermionic valves were not overshadowed by new companies devoted exclusively to the manufacture of semiconductors, but retained their pre-eminent position in the industry by taking up and developing the new product.

A factor lending immense force to the ability of the Japanese electronics industry to compete successfully with that of America and Europe is the method of financing employed, this being particularly advantageous in the semiconductor manufacturing sector with its rapid product cycles, and where fairly short-term fluctuations in sales regularly occur. A large proportion of the capital requirement of Japanese companies is supplied by bank loans. Abegglen points out that, with high debt to equity ratios, Japanese companies need not finance their growth out of retained earnings, but provided that they earn enough to cover the interest on their debt there is little financial constraint on their growth.[2] The commercial banks, who operate with almost all their deposits on loan, depend in turn upon the Bank of Japan, which controls the availability of funds. By this means the Bank of Japan—and ultimately the Government—controls the fortunes of individual companies. This policy allows favoured semiconductor companies to continue to operate whilst making considerable financial losses over a given period, a situation which would cause grave embarrassment to an American or European semiconductor manufacturer in a similar position. Consequently, unlike the situation in the United States, where the fortunes of individual companies have varied greatly over the last three decades, the major Japanese firms manufacturing semiconductors have remained financially strong since their entry into the field, This situation has also doubtless been assisted by the previously stated fact, that in addition to manufacturing semiconductors, they also produce a wide variety of other goods, these supplying markets which are not subject to the rapid trade cycles associated with semiconductor manufacture. For example, Fujitsu Ltd. produce not only semiconductors but telecommunication systems and equipment, and also data processing systems. Hitachi Ltd. manufacture, among other things, power generating plant, chemical plant and equipment, steelmaking equipment, rolling stock and road vehicles, elevators, compressors, pumps, chemical products, wires and cables, metal products, scientific instruments and electron tubes.[3] Furthermore, these companies are typical of the industry rather than being exceptional in this respect.

Unlike the situation in Europe, where American economic penetration in the form of capital investment into the industry has been significant, little progress has been made so far in this respect in the case of Japan. Mackintosh states in this context that: 'The major road block to a venture in Japan is the banking/ industrial relationship' and he points out that Japanese banks play a much stronger role in industry than in the United States even to the point of exerting pressure to remove managers if this is felt to be necessary.[4] Referring to this in the same article, he concludes:

> 'Therefore it is unlikely that a US company wanting to enter the Japanese market would agree to that kind of leverage in the hands of a Japanese bank.'

The major exception to date is Texas Instruments, who currently manufacture integrated circuits at their factory in Tokyo. Writing in 1982, Mackintosh suggested that the only way in which American companies might be able to obtain a foothold in the Japanese market would be via marketing agreements with local firms, or by joint ventures between Japanese and American semiconductor companies.

In common with the United States, the Japanese semiconductor industry possesses the advantage of defined skill clusters, which are situated in the area of Tokyo and also in the Osaka–Kobe region. The former district contains factories built by Fujitsu Ltd., Hitachi Ltd., the Mitsubishi Electric Corporation and the Toshiba Corporation, in addition to a number of smaller firms. The Osaka–Kobe region includes, amongst others, factories built by the Matsushita Electronics Corporation and the Toshiba Corporation. Because of the structure of the Japanese semiconductor industry, diffusion of knowledge does not depend on the existence of skill clusters in the same way as in the United States, where movement of personnel from one company to another is an important factor. It has, however, the advantage that highly important services are near at hand, and, since the major research and development facilities are also situated in these areas, close contact between research, development and production is facilitated. However, during the present decade a number of semiconductor firms have set up factories in the southern island of Kyushu in order to take advantage of the cheaper land and reduced labour costs in that region, suggesting that at some point these economic advantages outweigh the skill cluster element.[5]

Rather than set up fabrication facilities in South-East Asia, Japanese semiconductor manufacturers have concentrated on automating their production facilities to a greater extent than the Americans. As late as 1973, by which time US based companies were operating 128 offshore assembly plants, the Japanese semiconductor industry had only built four of these establishments.[6] A significant advantage of the Japanese strategy of automation is that it has resulted in the achievement of quality standards substantially higher than those attained by American semiconductor companies.[7]

By the end of the 1970s the leading Japanese semiconductor manufacturers began to show an interest in setting up fabrication facilities in both Europe and the United States. For example, by 1982, Fujitsu, Hitachi and NEC had established production facilities in Eire, W. Germany and Scotland, respectively, and by May 1984 Toshiba had opened a factory in West Germany to produce integrated circuits. NEC had planned in addition to construct a fully integrated $100m. 11 000 m^2 facility in the USA at Roseville, near Sacramento, California.[8] In 1981, Fujitsu announced the completion of an IC assembly plant in San Diego, California. Also by this time Hitachi had established a semiconductor facility at Irving, Texas, and Toshiba at Sunnyvale, California. In October 1986, Fujitsu announced its acquisition of Fairchild Semiconductor, based at Palo Alto, California, previously owned by the French firm Schlumberger Ltd. of Paris.[9] It is a matter of note that these ventures in the United States are centered in the existing skill clusters, thereby obtaining the advantages offered by that situation. As the result of these activities, Japanese manufacturers have steadily increased their share of the US semiconductor market. Although this amounted to only 2% in 1979,[10] B. Seducca, writing in 1985, mentions a current figure of 17%. Conversely, the American share of the Japanese semiconductor market at that time was 11%, a figure that had remained static for the previous decade.[11] Of the major US semiconductor manufacturers, in 1986 only Texas Instruments, Fairchild Semiconductors and Motorola have manufacturing plants in Japan.

A significant difference between the United States and Japan is that, unlike in America, where labour mobility is high and personal advancement may be obtained by changing companies, a Japanese employee, once recruited, is employed for life, his salary depending upon his age, education and length of service rather than by his position or productivity. Career-long training courses for engineers and scientists are universal among the semiconductor manufacturing companies, in addition to those offered by technical societies such as the Japan Union of Scientists and Engineers and the Japanese Institute of Standards. In contrast to the situation in the United States, where firms tend to encourage outside education, Japanese companies tend to favour in-house training. For example, the Toshiba company allows graduates one day a week to study an allied discipline. A mechanical engineer would, for instance, be given one day a week to study electronics, and an electronic engineer might be given, conversely, a day off to study mechanical engineering.[12] Certainly, a significant factor in Japan's industrial success has been the existence of a large highly educated workforce. For example, on a *per capita* basis, in 1977, Japan had almost three times as many electrical and electronic engineers as the United States, more than four times the total number in Great Britain, almost six times that of France and approximately 70% more than West Germany.[13]

Finally, one consequence of the move to higher levels of semiconductor device integration is that long-term planning has now become essential. This requires sustained team effort, and therefore a stable workforce. The high labour mobility of the American semiconductor industry, previously considered an advantage, may well be a handicap in this new situation.

It is therefore of great importance that the particular structure of Japanese industry makes long-term planning possible, and the success of the semiconductor industry in Japan is due in no small measure to the astute way in which this has been carried out, especially over the last two decades, under the leadership of the Ministry of International Trade and Industry (MITI).

Although state control does not exist in a socialist form, Abegglen describes the situation by quoting *The Economist* in the following words:

Table 7.1 Increase in Japanese semiconductor sales

All semiconductor sales ($ × 10³)			
Year	1956	1957	1958	1959
	2,192	10,786	24,664	53,316

'The ultimate responsibility for industrial planning, for deciding in which directions Japan's burgeoning industrial effort should try to go, and for fostering and protecting business as it moves in those directions, lies with the government.'[14]

7.1.2 Nature of the Japanese challenge to the dominance of the US semiconductor industry

Work on semiconductor devices under licence started in Japan in the early 1950s and the first transistor radio was manufactured in August 1955 (less than a year after the first transistor radio had been produced in the United States by Texas Instruments in October 1954). This was the 'Sony' TR 55 portable, which used five germanium junction transistors and two semiconductor diodes. It was the product of Tokyo Tsushin Kogyu Ltd., a licensee of Western Electric. (This company, now the Sony Corporation, is unusual in being a successful new Japanese semiconductor company.) Events moved rapidly, and by July 1956, five firms were licensees of American patents. They were Hitachi, Tokyo Tsushin Kogyu Ltd., Mitsubshi Electric Mfg., Tokyo Shibauro Electric and Kote Kogyo. By this time germanium N–P–N grown junction and also germanium N–P–N and P–N–P alloy junction devices were being manufactured and products included table model receivers, battery operated tape recorders, superheterodyne pocket receivers, regenerative pocket receivers and hearing aids.[15]

Semiconductor sales rapidly increased during the latter half of the 1950s as illustrated by figures published by the Japanese Electronic Industries Development Association (Table 7.1).[16]

In addition to this rapidly increasing number of devices produced, the shift from thermionic valves to transistors in domestic equipment was now well under way. For example, of five million radio receivers produced in 1958, more than half were transistorised portables. In 1959, ten million radio receivers were produced and eight million of these were transistorised.[17]

During the late 1950s transistor exports rapidly increased in number, and by the end of the decade a considerable proportion of these went to the United States, as shown by Table 7.2.

This first challenge to the dominance of the American semiconductor industry, which took place in the late 1950s, was entirely in the field of consumer electronics. Quantities of cheap germanium devices were then being exported to the United States in the form of both computers and equipment.

One of the most successful products was the cheap portable radio, as can be seen from figures published by *Electronics* (Table 7.3)

By 1959 the Japanese semiconductor industry had captured 50% of the American market for portable radios.[18] The predictable response was a demand for import controls, for instance, the *Electronics* issue of 6th January 1961 drew the conclusion:

> 'military requirements for transistors will expand tenfold in the next five years. Transistor imports of the magnitude of these from Japan could slow up the capacity essential for defence. Transistor imports also could reduce industrial outlays for R & D.'[19]

The article continues

> 'we do believe the time has come for Federal authority to consider reasonable import control, such as establishing quotas on imports from countries where labour costs are low and transportation costs are not a limiting factor.'

Nevertheless, these imports amounted in 1959 to only 0·4% (by value) of the total US semiconductor home market,[20] and seen within this context, it would appear that fears within the US industry at the time were somewhat over-emphasised.

One response by American semiconductor companies to the Japanese inroads into their home market was to set up offshore assembly plants, the first of these being completed in 1963 by Fairchild. This plant was situated in Hong Kong, where labour costs were relatively low. By 1974, 56 of these units had been set up in the Far East, in addition to eight fabrication plants, six of which were in Japan.[21] Finan states that, by 1972, a total of about 89 000 people were employed in offshore facilities, compared with approximately 85 000 people in the United States.[22] The result of this event was effectively to change the American semiconductor manufacturing industry from a predominantly home based endeavour into an international operation. Although factors other than the Japanese successes undoubtedly assisted in bringing about this shift in capital, it is significant that it took place precisely when the challenge from overseas made itself felt.

A new development of a technical nature was however soon to make itself felt, to the disadvantage of the Japanese semiconductor industry. At the beginning of the 1960s most transistors produced in that country were of germanium type. Since Japan did not have a defence industry, and the American Government

Table 7.2 Transistor exports (no. of devices)

Year	1957	1958	1959
Total exports	11 187	351 508	4 741 483
Exports to the USA	N/A	10 020	2 393 365

Source: Japanese Finance Ministry.

Table 7.3 Increase in imports of radios from Japan to USA

Year	1956	1957	1959
Radio imports from Japan to the USA ($ million)	2·5	5·6	55·0

Source: *Electronics* 'Market Data Survey', Jan. 1971

does not issue Defense Contracts to foreign countries, there had been little incentive for Japan to enter the silicon market. At this time, approximately twice as many germanium devices as silicon were being manufactured in the United States. However, by the end of the decade, the situation in the latter country had completely changed, and silicon devices, with their better operating characteristics, had almost completely replaced their germanium equivalents (see Fig. 7.1). If the manufacture of integrated circuits were also included, the shift from germanium to silicon during this time is even more significant.

With the considerable drop in price of silicon devices which took place in the early 1960s, Japan's predominance in germanium device production was rapidly undermined, placing their semiconductor industry at a considerable disadvantage compared with that of the United States. This situation is illustrated by Fig. 7.2, which shows that by 1965 the average unit cost of discrete silicon devices had fallen to almost that of their germanium equivalent. Furthermore, since it was technically impossible to manufacture germanium planar transistors or integrated circuits, the Japanese semiconductor industry was forced at this time to completely review its strategy. In order to stay in the

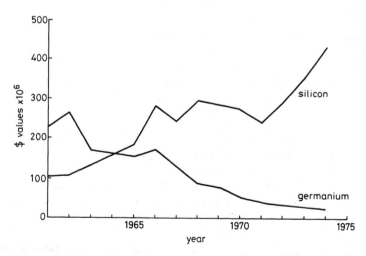

Fig. 7.1 US silicon bipolar sales and germanium bipolar sales (Source: *Electronics* Yearly Market Data Survey)

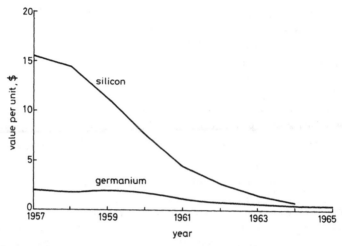

Fig. 7.2 Value of germanium and silicon devices per unit (Source: EIA Market Data Book, Washington, DC, 1974, p. 57)

semiconductor business, it was necessary to change to the manufacture of silicon devices and to catch up with American silicon-based technology.

A further development acting to the disadvantage of Japanese semiconductor firms was that an increasing mutual interdependence had arisen in the United States since the early 1960s between the computer industry and semiconductor manufacture, and this affected commercial, industrial and military requirements. This was a growing long-term trend, and by 1978 56% of the American consumption of semiconductors was in the field of computer/electronic data processing and office equipment sector, the consumer sector of the market accounting for only 9% of consumption (by value). This situation contrasted sharply with that of Japan where, even at this late stage, 63% of semiconductor consumption (by value) took place in the consumer sector.[23]

Another disadvantage involving the production of low cost germanium devices for the consumer market was that salaries in Japan had risen by about 10% per annum between 1960 and 1965, with a 13% jump in the latter year. At this time, a production worker would start at about $168 per month, including fringe benefits, compared with about $15 per month in Taiwan and $30 per month in Hong Kong, this changed situation offering a considerable advantage in labour costs to American semiconductor assembly facilities in the latter countries, which were now being rapidly set up.[24]

A consequence of the growth of digital computer systems both in number and complexity was a shift in circuit design within the electronics industry away from analogue techniques, and this in turn has led to a shift in demand from bipolar to field effect transistors and also from discrete devices to integrated circuits. These shifts in emphasis have been particularly evident in both industrial and military applications. At the beginning of the present decade, the US semiconductor industry led both that of Japan and Europe in the above mentioned trend, as can be seen from figures compiled for 1981 (Table 7.4).[25]

It can also be seen from Table 7.4 that by this time the Japanese were world leaders in the discrete device market (and consequently in the world consumer market). The main strength of the American semiconductor industry now lay in the more advanced technology of silicon integrated circuits. This situation reflected to a great extent the sophisticated demands of the military and industrial complex and the ability of the US semiconductor manufacturers to effectively supply it.

Work on integrated circuits in Japan started in 1964, and by the autumn of 1965, Fujitsu Ltd., Hitachi Ltd. and the Nippon Electric Company, the 'big three' of the Japanese computer industry, were ready with computer using IC's. Five other semiconductor companies had also by this time developed mono-lithic integrated circuits, diode transistor logic (DTL) being the most favoured approach to computer design. Research grants were being made available at this time by MITI for development of special ICs.

By 1965, the Nippon Electric Company, Hitachi, Fujitsu, Mitsubishi, Oki, Toshiba, Sony and Matsushita were all making integrated circuitry. It was at this time that long-term plans were being made which were to change the character of the Japanese semiconductor industry from a high-volume producer of cheap commercial devices to that of a manufacturer of integrated circuits which was two decades later to dominate the field of very large scale integration. Ichano Isaka, Chief Engineer of the Electrical Industry Association of Japan, writing at this time, says: 'clearly the biggest change is the new accent on research. One of the greatest incentives to this approach is the desire to become independent of US patents'. Certainly one problem resulting from a technical lag is the expense incurred in paying royalties to the owners of patents and in this respect the Japanese were at that time at a clear disadvantage in the field of integrated circuitry. Indeed, taking the semiconductor industry as a whole, during the 1970s Japanese semiconductor firms were paying approxima-tely 10% of sales in royalties to American companies. As late as 1973, Mackintosh was able to write that:

> 'the Japanese semiconductor industry lags, technically and commer-cially, well behind the US, as evidenced by the preponderance of US

Table 7.4 Comparison of international semiconductor production

Production of discrete devices (percentage of world production in terms of value)	Japan	40%
	USA	31%
	Europe	21%
	Rest of world	8%
Production of integrated circuits (percentage of world production in terms of value)	USA	48%
	Japan	33%
	Europe	15%
	Rest of world	4%

made MOS ICs currently used in Japanese electronic calculators and similar equipments'.[26]

However, the picture of American technical predominance began to change significantly by the late 1970s, largely due to the effects of the long-term strategy pursued by the Japanese government and semiconductor industry, under the aegis of the 1971 programme, initiated and co-ordinated by the Ministry of International Trade and Industry (MITI), with its emphasis upon large-scale integrated circuit development and fabrication. An ensuing programme concentrating on very large-scale integration followed in 1976. Exact figures regarding investment in these programmes are not available, but the amount of money involved was substantial. Braun and Macdonald estimate that, between 1971 and 1977, MITI invested about $100 million in a large scale integration programme and between 1976 and 1979 about $150 million in a very large-scale integration programme.[27] Mackintosh, writing in 1978, rates investment in the very large scale integration programme at the somewhat higher figure of $65 million per annum.[28]

The result of these heavily financed and ambitious programmes has been twofold. Firstly, the attainment of a significant technical lead over the United States in large-scale memories has currently been achieved, and secondly, resulting from this lead, the virtual domination of the world-wide very large-scale integration market by Japanese products. By 1978, Fujitsu had produced a 64 Kbit random access memory, thus closing the technology gap previously existing in that field, and since then efforts have been intensified. Bell, writing in the *IEEE Spectrum* in April 1986, states that during 1983–84, Japanese capital equipment investment for semiconductor manufacturing was 1·6 times that of the United States (for example, $4·7 billion compared with $3·1 billion).[29] At the time of writing, Japan leads in sales of all large memories, specifically 64 Kbit, 256 Kbit and 1 Mbit dynamic random access memories (DRAM's); also in emitter-coupled logic (ECL) random access memories (RAM's) and MOS arrays. Bell estimates that currently (1986) the Japanese manufacturers are 'two years down the cost experience curve' compared with US competitors.[30]

Certainly American firms have come under considerable pressure in this particular area. By 1983, Japan had surpassed the United States in its world market for DRAM's.[31] Wallish, writing in the *IEEE Spectrum* in April 1986, stated that:

'The Japanese share of the memory market has gone from almost nothing in the late 1970s to more than half the market for 64 K RAMs, about 90% of the market for 256 K RAMs and virtually the entire market so far for 1 Mbit RAMs.'[32]

Writing in the *Financial Times* (December 1985), B. Seducca pointed out that only one US manufacturer of 1 Mbit memories remained in the field (Texas Instruments).[33] The effect of this achievement has been to force the United States semiconductor industry into a highly defensive posture. For example, the editorial in *Electronics*, published in March 1982, states that:

'It is now apparent to American electronics firms that they cannot continue to do battle with the Japanese as individuals, as they have

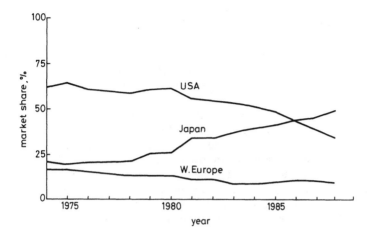

Fig. 7.3 Estimated share of world semiconductor market (Source: Reference 6, p. 121; and *Electronics Times*, 7 Aug. 1986, p. 1

against each other. Until now the electronics industry have gotten only nominal help from the US Government in neutralising the economic advantage ensured by the Japanese electronics industry.'

The article goes on to quote N.C. Norris, Chairman of Control Data Corporation, who proposed:

'the formation of a new R & D consortium of semiconductor and computer manufacturers that would speed and broaden the development of advanced technology and combat the Japanese development programmes already under way.'

Considerable pressure from the American semiconductor industry has lately resulted in decisive government action. In February 1986 the US Department of Commerce imposed a heavy import surcharge on a range of Japanese semiconductor products including 256 Kbit and 1 Mbit memories. P. Swart and S. Parry, writing in *Electronic Times* state that:

'EPROMs selling in the US for $4 will now have a $7·52 duty levied on each chip. A DRAM selling for $2 will now have $2·18 worth of duty added.'[34]

However, in spite of these measures, in 1988 the Japanese share of the US semiconductor market amounted to 20%, and by the beginning of the following year 90% of the world DRAM market had been secured. Some indication of the overall effectiveness of the Japanese challenge may be obtained from a consideration of Table 7.5, which indicates the extent to which the leading American semiconductor companies have been displaced in ranking as the major producers of semiconductor devices during the period 1978 to 1985. Fig. 7.3 reinforces this picture, indicating the extent to which the Japanese share of the world market has grown since the mid-1970s, so that now, a decade later, it has risen to a level surpassing that of the United States.

The situation outlined in Table 7.5 has arisen within a market which continues to expand rapidly, as shown by Table 7.6. It can be seen that the current average growth rate of semiconductor production in the case of Japanese firms is significantly higher than for the corresponding American companies.

Capital investment in chip production continues at a very high level, amounting to an estimated $5 billion for 1989.

This entirely new situation brings into question not only the future viability of the American semiconductor industry, but consequently raises the question as to whether, to quote Kehoe, 'the USA will become reliant upon foreign supplies for its most strategically significant technology and supplies'[35], or, alternatively, by the application of selective tariffs and import surcharges ensure that a viable semiconductor industry remains to supply nationally vital needs.

7.2 South Korean semiconductor industry

South Korea is unusual in that, alone amongst the nations of Eastern Asia (excluding Japan), it has developed a viable indigenous semiconductor industry.

Originally, South Korea played a fairly minor role as an offshore assembly area for semiconductors, only nine offshore assembly plants owned by the United States being in operation by 1974.[36] No fabrication plants were in existence at that time.

By 1980, monthly wages for semiskilled electronics workers (including fringe benefits and bonuses) had risen to an average of $240 US compared with $100 US in the Philippines, $140 US in Taiwan, and $180 US in Malaysia.[37] This relative increase in wages in Korea, compared with that in other countries in Eastern Asia, was cited by the Government-formed Korean Institute of Electronics Technology (KIET) as the reason for local integrated circuit assembly operations becoming progressively less profitable, and the prospect of the collapse of the industry could not be discounted.[38] In order to retrieve the situation, the Korean government in 1979 allocated approximately $60m., including about $29m. in World Bank loans, for research, development and pilot production of integrated circuits.[39]

By 1982, the following companies were engaged in semiconductor manufacture: Gold Star, at Gyeongsang, Anam Industrial Co. Ltd. at Inchon, and the Samsung Semiconductor and Telecommunications Co. Ltd., at Kyuggi-Do, the latter company manufacturing CMOS, PMOS, and bipolar integrated circuits in addition to some discrete devices.[40] In May 1984, Samsung (a conglomerate on Japanese lines producing a tenth of South Korea's gross national product) built a very large integrated circuit plant in the suburbs of Seoul, and only five months after production in August 1984 had reached an output of six million piece parts per month by January 1985.Samsung established a Californian base for semiconductors in the late 1970s, and Tristar Semiconductor was formally

Table 7.5 Semiconductor sales (including ICs) of the ten major world semiconductor manufacturers in order of ranking

Year / Rank	1978	1981	1984	1985	1987	1988
1	TI	TI	TI	*NEC*	*NEC*	*NEC*
2	Motorola	Motorola	*NEC*	*Hitachi*	*Toshiba*	*Toshiba*
3	Philips	*NEC*	Motorola	TI	*Hitachi*	*Hitachi*
4	*NEC*	Philips	*Hitachi*	Motorola	Motorola	Motorola
5	*Hitachi*	Nat. Semicon	*Toshiba*	*Toshiba*	TI	TI
6	Nat. Semicon	*Hitachi*	Nat. Semicon	*Fujitsu*	*Fujitsu*	Intel
7	*Toshiba*	*Toshiba*	Intel	Philips	Philips	*Matsushita*
8	Fairchild	Intel	*Fujitsu*	Nat. Semicon	Intel	*Fujitsu*
9	Intel	Fairchild	Philips	*Matsu-shita*	*Mitsu-bishi*	Philips
10	Siemens	*Matsushita*	*Matsushita*	Intel	*Matsu-shita*	*Mitsubishi*

Source: 'Profile of world-wide semiconductor industry' (Benn Electronics, 1982) *Financial Times* Survey 'Semiconductor industry' 6 December 1985 'Dataquest', 1988, 1989
Note: Japanese firms in italics.

Table 7.6 Current growth rate of the ten major world semiconductor manufacturers

Year	1986	1985 Production ($m)	Estimated 1986 production	Estimated % increase in production
Rank				
1	*NEC*	2070	2825	26·7
2	*Hitachi*	1875	2045	8·3
3	*Toshiba*	1460	2005	27·2
4	Motorola	1800	2000	10·0
5	TI	1830	1830	Nil
6	*Fujitsu*	1055	1550	31·9
7	Philips	1010	1215	16·8
8	*Matsushita*	853	1165	26·8
9	Nat. Semicon.	950	970	2·0
10	*Mitsubishi*	600	910	34·0

Source: *Electronics Times*, 8 January 1987, p. 1
Note: Japanese firms in italics.

established as a subsidiary in 1982. Renamed Samsung Semiconductor, this plant now employs about 1000 personnel. Samsung are currently considering setting up a similar semiconductor operation in Europe.[41]

It is the oldest of Korea's conglomerates, having entered the semiconductor business in 1974. This company has the advantage of having a strong commitment to research and development, and before launching into large scale integration had produced both discrete devices and medium scale integrated circuitry. In 1987, Samsung was the only company making a VLSI semiconductor to its own design (a 250 K DRAM chip) and in addition had by the end of that year completed the design of a 1 Mbit DRAM. Capital investment by Samsung in 1986 amounted to $500 million and sales in that year were approximately $300 million. Samsung's wafer capacity then amounted to 97 000 per month, making it the largest semiconductor manufacturer in Korea.

Another large conglomerate, Hyundai, entered the business of semiconductor manufacture in 1983. In 1985, Hyundai signed a licensing agreement with Vitelic Corp., San José, California, to produce 256 K DRAMs,[42] under a contract with Texas Instruments, at its plant in Ichon. Problems, however, arose at this time owing to a lack of technical expertise in device fabrication. Also, investment in research and development has been criticised as being inadequate.

Government plans in 1984 included increasing the number of chip manufacturers to six, and also to establish an integrated circuit and software design centre.[43] In spite of heavy financial investment and encouragement from government, the South Korean semiconductor industry was still far from being financially viable by the end of 1987, sales amounting to only approximately one fifth of total investment by the four major companies then in operation (Lucky–Goldstor, Samsung, Hyundai and Daewoo).

It is interesting to note that this government sponsored attempt to build up a significant local semiconductor manufacturing industry has much in common with the Japanese approach. For example, a large injection of capital has been made into an industry comprising a small number of vertically integrated electronics companies geographically situated within a minor skill cluster in the region of Seoul. However, lacking the advantage of a strong home base, as in the case of Japan and the United States, the Korean semiconductor industry will hope to concentrate almost entirely on selling to overseas countries, and this cannot fail to be a disadvantage, rendering it particularly prone to international market fluctuations.

It is also of interest that this new local industry has arisen in response to increasing wage rates thereby threatening the existence of the foreign owned semiconductor assembly facilities then in operation. Only time will decide whether this development is an isolated event, or the beginning of a trend in which indigenous local semiconductor industries will spring up within the Far East as wage rates generally rise.

7.3 References

1 ABEGGLEN, J.C.: 'The economic growth of Japan', *Scientific American*, March 1970, p. 31
2 *Ibid.*, p. 33
3 'Electronics in Japan', *Electronics Association of Japan*, pp. 130–131
4 MACKINTOSH, I.M.: 'Profile of the world-wide semiconductor industry', (Benn Electronics, 1982), pp. 100–101
5 Conversation with K. Kitagawa, Engineer, Toshiba Ltd.
6 BRAUN, E. and MACDONALD, S.: 'Revolution in miniature' (CUP 2nd Ed., 1982) p. 150
7 GREGORY, G.: 'Japanese electronics industry: enterprise and innovation', (Japan Times Ltd., Tokyo, 1985)
8 MACKINTOSH, I.M.: *op.cit.*, p. 87
9 *IEEE Spectrum*, Jan. 1987, p. 45
10 GREGORY, G.: *op.cit.*, p. 197
11 SEDUCCA, B.: 'String of poor results for big companies', *Financial Times Survey*, 6 December 1985
12 Conversation with K. Kitagawa, Engineer, Toshiba Ltd.
13 GREGORY, G.: *op.cit.*
14 ABEGGLEN, J.C.: *op.cit.*, p. 33
15 *Electronics*, July 1956, p. 120
16 *Electronics*, 27 May 1960, p. 66
17 *Ibid.*
18 *Electronics* 'market data survey', Jan. 1971
19 *Electronics* 'market data survey', 6th Jan. 1961, p. 94
20 *US Dept. of Commerce*: 'Semiconductors': 'US Production and Trade', (US Government Printing Office, Washington DC, 1961, pp. 15–16
21 US Dept. of Commerce: 'A report on the semiconductor industry', (US Government Printing Office, Washington DC, 1979) p. 55–56
22 FINAN, W.: 'The international transfer of semiconductor technology through US based firms' (National Bureau of Economic Research, NY, 1975), pp. 50–52
23 RUDENBURG, G.: 'World semiconductor industry in transition: 1978–83', p. 11
24 *Electronics*, 13 December 1965, p. 77
25 MACKINTOSH, I.M.: *op cit.*, p. 17

26 MACKINTOSH, I.M.: 'Dominant trends affecting the future structure of the semiconductor industry', Radio and Electrical Engineer, 1973, **43**, (1) & (2), p. 150
27 BRAUN, E. and MACDONALD, S.: *op.cit.*, p. 150
28 MACKINTOSH, I.M.: 'A prognosis of the impending intercontinental LSI battle', *Microelectronics J.*, 9, 1978, p. 68
29 BELL: 'A source for solid state', *IEEE Spectrum*, April 1986, pp. 71–74
30 *Ibid.*
31 'Assessing Japan's role in telecommunications', *IEEE Spectrum*, June 1986, p. 51
32 WALLICH, P.: 'US semiconductor industry: getting it together', *IEEE Spectrum*, April 1986, p. 77–78
33 SEDUCCA B.: 'String of poor results from big companies', *Financial Times*, 6 December 1985
34 SWART, P. and PARRY, S.: 'Japan to pay the price for dumping chips in US', *Electronics Times*, 20 March 1986
35 KEHOE, L.: 'US chipmakers face extinction', *Financial Times*, 6 December 1985
36 'A Report on the semiconductor industry', (US Department of Commerce), *Ibid.*, pp. 85–86
37 MACKINTOSH, I.M.: 'Profile of the worldwide semiconductor industry', (Benn Publications, 1982), p. 74
38 *Ibid.* p. 77
39 *Ibid.* p. 76
40 *Ibid.* p. 77
41 McLEAN, M.: 'The Koreans are coming', *Electronics Times*, 11 December 1986, p. 20
42 SEDUCCA, B.: 'Aggressive drive into mass markets', *Financial Times Report*, 6 December 1985
43 *Ibid.*

Chapter 8

Review of the European semiconductor industry

8.1 General review

The Western European semiconductor industry forms the third major grouping in the world, following that of the United States and Japan in size. Several important differences exist between this industry and its two major competitors. Firstly, it is fragmented into a number of national groups, does not possess any recognisable skill clusters, and its markets have made little headway overseas. Furthermore, its home market has been strongly and increasingly penetrated by both the Americans and (more recently) the Japanese, who not only export a large number of devices (both individually and within assembled equipment) to Europe, but have, in addition, built semiconductor fabrication facilities in all the major Western European countries. In 1986, American semiconductor companies held approximately 45% of the European semiconductor market by value, and the Japanese about 11%.[1]

Perhaps the most important disadvantage suffered by all European semiconductor manufacturing countries is that no major source of finance has been forthcoming to generate and sustain the industry. Unlike the situation in United States, no large military contracts have been made available, enabling economically profitable production to take place during the initial phase of manufacture. For example, in 1972 the military market in Western Europe amounted to only 14% of total semiconductor sales compared with 24% of the much larger American market.[2] Neither, unlike the Japanese, has any Western European government initiated and sustained any long-term plan for the industry. Consequently, semiconductor firms in Europe, which tend (like those in Japan) to be vertically integrated, have had to a great extent to finance their semiconductor manufacturing operations out of company profits. Financial aid from individual governments, often of quite a substantial nature, has been made, although largely to research institutions, both public and private, but what is really significant is that in general the critical production–development area has been relatively neglected in this respect.[3]

With populations equal to about half that of Japan, the major Western European countries, Britain, France, W. Germany and Italy, all possess considerably smaller home markets, and it is highly doubtful if, purely on size alone, these would be economically viable if considered individually. Some recognition of this situation has latterly resulted in links being established between various Western European semiconductor firms.

Although in certain areas individual European semiconductor companies have succeeded in producing highly advanced products, they have in general been unable to supply these in sufficient quantity at a competitive price, and it appears that the main difficulty lies in the production–development area. As semiconductor devices become more sophisticated, the capital cost of production equipment has risen, resulting in ever greater financial demands on individual companies. In common with semiconductor production in the United States and Japan, the effect of automation has been the employment of a smaller but more highly skilled labour force.

In order to obtain the advantage of current American technology, reduce the all important time lag between America and Europe in the diffusion of technical knowledge, and to gain access to the American semiconductor market, a number of European transistor manufacturers have acquired or obtained an interest in American semiconductor firms, examples being the acquisition of Signetics by Philips (Holland) in 1975, Fairchild by Schlumberger (France) in 1979, and Microwave Semiconductors by Siemens (Germany) in the same year. It should be noted that these acquisitions serve, to an important extent, a different purpose to the semiconductor fabrication facilities constructed in Western Europe by the American industry, the aim in the latter case being specifically to penetrate economically the European market, taking advantage of the reduced labour costs and (in some cases) financial assistance from local governments. This policy has also been followed more recently by the Japanese [namely Fujitsu (1989) and NEC (1987)] who have set up semiconductor plants in Britain.

The major producers of semiconductors in Western Europe are currently West Germany, the United Kingdom, France and Italy. Each of these countries will be considered in turn. In addition to the forementioned States, research and development, together with device production is at present being carried out in a number of other countries, principally in the Netherlands, but to a limited extent in Sweden, Switzerland, Eire, Austria and Spain.

Table 8.1 gives an indication of the relative size of the major European semiconductor companies currently engaged in integrated circuit production. Note that the figures quoted do not include production by overseas subsidiaries of European companies, and that American subsidiaries operating in Europe have also been excluded.

Within Europe itself, major changes in the relative strengths of the national semiconductor manufacturing industries have taken place. These changes are outlined in the following Table 8.2

8.2 British semiconductor industry

This section begins with a general review of the characteristics of the British transistor manufacturing industry (including the Mullard organisation) and considers various factors which appear to have contributed to its disappointing performance internationally. Because of the generally poor showing of

Table 8.1 Major semiconductor companies currently engaged in integrated circuit production

Country of origin	Company	IC production	
		$ million	
		1983	1986 (est.)
Netherlands	Philips	250	549
Germany	Siemens–AG	200	192
Italy	SGS–ATES	175	192
*	ITT Corp.	100	115
Britain	Ferranti Ltd.	75	—
France	Thompson–CSF	60	198
Britain	INMOS Ltd.	57	—
Britain	Plessey Co.	50	100
Germany	AEG Telefunken	40	52
—	Others	43	—

* ITT is a European-based American company.
Source: IC Engineering Corp. (US.) (1984)
       'Dataquest', *Electronics Weekly*, 28 Jan. 1987
       Computer Guardian, 30 April 1987

European semiconductor firms, and in particular those of Britain, it is likely that problems which have arisen within the British semiconductor industry will also have some relevance across a wider, European dimension. In this section,

Table 8.2 Semiconductor output

Country	1965	1984	1986	1987 (projected)
Germany	47	1290	1700	1910
United Kingdom	208	1110	1040	1150
France	59	670	810	920
Italy	25	450	500	575

Source: 'Gaps in technology'. OECD Report, Paris 1968
       Also, Motorola Marketing Research publication
       Electronics Times, 13 November 1986

therefore, an attempt has been made to identify the factors which have led to this poor performance, in order to assist in drawing general conclusions about the development of the European semiconductor industry.

This review is followed by a profile tracing the development of three of the major British semiconductor companies, and in addition the Mullard company, which must also be considered because of its special position as the major European semiconductor manufacturer.

8.2.1 Review of characteristics

A number of firms in the United Kingdom have been engaged in the manufacture of semiconductors since the early 1950s, the most important being Plessey, GEC and Ferranti, in addition to Mullard (a subsidiary of Philips), which, although controlled from Holland, possesses considerable local autonomy.

The size of the British semiconductor industry is relatively small. For example, in the years 1977–79, total sales by the three major British firms mentioned above amounted to only 6% of the semiconductor sales of Texas Instruments Inc.[4]

The pattern followed in the United States, where thermionic valve manufacturers were largely displaced as the leading transistor manufacturers by new companies entirely dedicated to the production of semiconductors, has not occurred in Britain. Instead, local subsidiaries of these (largely American) companies have successfully established themselves. For example, Texas Instruments (Bedford) Ltd., set up in 1957, had by 1968 displaced the Mullard–GEC merger ASM as the leading semiconductor manufacturer in the United Kingdom, taking 23% of the market compared with 22% for ASM.[5]

It is of interest to consider the possible reasons which have been advanced as contributing to Britain's relative lack of success in semiconductor manufacture and marketing compared with the corresponding industries in both America and Japan. It appears likely that the following factors have played a significant part, although the relative importance of each is difficult, if not impossible, to assess.

Owing to the technological lag existing between the US semiconductor industry and that of the United Kingdom, amounting to approximately two years during the 1950s and 1960s,[6] and which still appears to exist, efforts to break into the large-scale international semiconductor market have been extremely costly, owing not only to the problem of high unit costs during the early part of a production cycle, but also because of royalty payments for patents on newly developed processes. A highly significant handicap has been that, although American firms are usually prepared to grant licences on patents, obtaining the corresponding 'know-how' is much more difficult, although obviously this disadvantage does not apply to the American-owned local subsidiaries. (Braun and Macdonald state in this context that 'a survey of 42 American semiconductor firms revealed that only four companies regularly divulged process technology to foreign firms').[7]

Unlike the situation in the United States, where diffusion of 'know-how' between individuals within a particular skill cluster may make information fairly readily available, there is little labour mobility within the British semiconductor industry, either at a production level or in research and development, and any movement between British and American firms has usually been in the form of a 'brain drain' from research and developments departments in Britain to those in America.[8]

An important advantage possessed by foreign local subsidiaries over the indigenous industry is that research and development, information and also 'state of the art know-how', including production–development expertise may be rapidly transmitted by recruitment of personnel from the parent company, enabling new types of device to be put into production much more rapidly than otherwise would be the case. Advanced production equipment, already tried and tested by the parent company can be imported, saving valuable time and avoiding costly mistakes in buying (or manufacturing) unsuitable equipment.

Although considerable sums of money have been invested by Government in research and development, the vital area of development–production has been neglected in this respect and it is precisely in this latter area that the American and Japanese semiconductor industry has been particularly successful. It is significant that both the latter countries have invested considerable monetary support in production technology. It has been suggested that lack of success in getting devices rapidly into quality production may, however, not only be due to lack of financial investment, but may also be influenced for instance, by the relatively low status accorded to production engineering in Britain.[9]

A further factor contributing to Britain's lack of success in the semiconductor field may be that due to frequent changes in government and consequently government policy, any long-term industrial planning, such as instituted in Japan by the Ministry of International Trade and Industry, has been absent. The role of the commercial banks in Britain, geared to a 'seed time to harvest' mentality, has ensured that long term loans to industry from this source (of particular importance in the case of the semiconductor industry, subjected as it is to long-term periods of financial loss) have not been forthcoming.[10]

Unlike its American counterpart, the British defence industry is too small to finance contracts on such a scale as to enable large-scale production runs of devices to be initiated, and to subsidise these devices by providing a market during the early production period when unit costs are high. Access to American defence industry contracts have, of course, been excluded on grounds of national security. A major source of finance has thus been denied to the British semiconductor industry,[11] and this has perhaps been the single largest factor in ensuring the rapid development of the corresponding American industry. Certainly the importance of this cannot be too strongly emphasised. In view of this fact, it is perhaps surprising that American semiconductor companies operating within the United Kingdom have received favourable treatment in the form of British government research and development contracts through the Inter-services Committee for the Co-ordination of Valve Development (CVD)[12] and also qualified for financial support under the Government microelectronic support scheme.[13]

It is instructive to consider the reasons for the lack of success of spin-off semiconductor companies within the United Kingdom, since in so doing it is possible to indicate a further number of difficulties facing the semiconductor industry.

Unlike the United States, where the incidence of spin-off companies has been high, hardly any similar companies have emerged in Britain, and the few that have appeared have had little success. Golding, writing in 1972, lists only three spin-off companies, none of which have stayed the course.[14] For example, one of these, Microwave Semiconductor Devices (MSD) was set up in 1961 by four engineers form TI Bedford, under the sponsorship of Microwave Associates, an American semiconductor company. MSD began to manufacture silicon epitaxial planar transistors on a small scale at the beginning of 1962, but production ceased after a few months.

One important factor accounting for this lack of success is that the American lead of about two years in device technology means that by the time the spin-off company in the UK gets into production, its market rivals in the United States will be selling similar devices at a much lower unit cost, and in order to compete effectively it is necessary for the spin-off company to sell at a financial loss for perhaps a long period until their unit costs approach those of their competitors.[15] To survive this financially stringent period requires considerable financial backing, and it is perhaps not surprising that this money has not been forthcoming. In contrast to this state of affairs, spin-off companies in the United States are not faced with the problem of technical lag, and, as previously mentioned, have access to Government contracts denied to their British rivals. Regarding this second factor, Golding writes:

> 'It is the absence of a military requirement on the American pattern which renders cost of entry prohibitive for early innovative spin-offs in the UK.'

It should be remembered, however, that these pressures do not only apply to spin-off companies, but also to existing semiconductor concerns within Britain, although the problem then presents itself in a somewhat attenuated form owing to their larger financial resources. The result has been that, although such firms as Ferranti, Plessey and GEC have not been put out of the semiconductor business by their American rivals, their performance has certainly been considerably curtailed. A third factor mitigating against the emergence of spin-off companies in the UK is that as device manufacture becomes more complex, the capital cost of production equipment has rapidly increased, inhibiting the launching of new ventures. Yet a further disadvantage facing spin-off companies in Britain is that the number of semiconductor engineers trained to an adequate level in the latest technology is small compared with that of the United States (and possibly also Japan) making the recruitment of an efficient production force correspondingly more difficult. Added to this, the absence of skill clusters means that this potential labour force is more widely spread and the problem of re-location of labour correspondingly greater.

In view of all these disadvantages, it appears impossible to set up successfully a completely new semiconductor manufacturing facility in the United Kingdom without considerable financial investment, and it is extremely doubtful if this could be done without Government support on a very large scale.

Nevertheless, in addition to providing economic assistance to the indigenous semiconductor industry, the alternative strategy of financing an entirely new semiconductor facility has been attempted. In 1978, the Labour Government, through the National Enterprise Board, initiated a project to set up a new British semiconductor company under the name of Inmos, to manufacture advanced integrated circuits. An initial $50 million was made available for early development to be followed at a later date by a further $100 million in two instalments for building production plant and also for research and development.[16] Headed by Dr. R. Petritz, founder of the Mostek Corporation, it has headquarters at Colorado Springs, USA, and at Bristol, UK, with its British production plant situated in South Wales. Following a change of Government in 1979 to the Conservatives, an unsuccessful attempt was made to sell Inmos to British private companies. Inmos has not been a financial success. It was purchased in 1984 by Thorn EMI for the sum of $95 million but has continued to lose money. What was obviously intended to be an attempt to take advantage of American expertise in both technology and management in order to establish a viable centre for advanced semiconductor manufacture in the United Kingdom does not really seem to have so far succeeded, partly because of the present government's commitment to an alternative economic strategy, and also because in order to achieve a real breakthrough, investment on an even larger scale is needed.

Finally, an important factor contributing to European lack of performance in the semiconductor field, and perhaps applying to Britain in particular, has been lack of a successful management and marketing strategy. This has been recognised by overseas competitors and has been the subject of some discussion and analysis.[17] Certainly in the United States, the status, quality and renumeration of sales engineers has been relatively greater than is generally the case in Europe. In the highly technical field of semiconductor device engineering, it is important that the salesman is in a position to identify the customers' needs, and assist in product development. British and European device manufacturing companies, being vertically integrated in structure, have tended to inherit senior management personnel from outside the semiconductor field. Consequent problems of adjustment and failure to appreciate the potential of the new technology must be regarded as highly likely in this situation.

8.2.2 Mullard Ltd.

This company was founded in September 1920, and became a wholly owned subsidiary of Philips in 1927. During the Second World War, unlike its parent company in Holland, it avoided German occupation, and certainly for some years after that event exercised a degree of autonomy within the Philips organisation. However, from the 1960s onwards, control has tended to become centralised and close contacts are now maintained with Holland and also with other Philips subsidiaries.

When Mullard entered the field of semiconductor manufacture it was as a highly successful large-scale producer of thermionic valves. The main strength of the company then lay in the area of high volume mass-production, and it had become a market leader in this field.

Research and Development began at Redhill in Surrey in 1954, and manufacture of glass-encapsulated germanium diodes and germanium alloy transistors began the following year at the Mitcham factory. (Mullard were assisted in this initial work by the parent company, Philips, who actually produced a working transistor within a week of the announcement by Bell in 1948, their sole source of information being the American daily papers.)[18] In 1957, a large manufacturing plant was completed at Millbrook, Southampton, and transistor production was transferred from Mitcham. Large-scale production of germanium alloy transistors at that location was successful and Mullard were soon able to capture the major part of the British semiconductor market, their share amounting to 55% in 1958.[19]

Fabrication of planar devices on a mass production basis did not begin until 1966, when germanium sales were falling off rapidly. Mullard at this time continued to pursue its policy of making device fabrication equipment, and time and money were certainly lost in this exercise before it was decided to buy from specialist manufacturers in the United States. Certainly at this time Mullard was at least two years behind TI Bedford in the technology of planar manufacture. This delay in getting into production resulted in a dramatic fall in the Mullard share of the semiconductor market.[20] Although they have not regained their earlier position as the major British semiconductor manufacturer, they currently claim to produce the widest range of integrated circuits in the United Kingdom for industrial, consumer and military purposes. They are a vertically integrated company, supplying the domestic, industrial, scientific, medical and military markets. In addition to Mitcham and Southampton, factories are in operation at Simonstone and Blackburn in Lancashire, Tyne and Wear, and Hazel Grove in Cheshire, and manufacture a wide variety of electrical goods. By the mid 1980s Mullard employed 12 000 people, nearly 1000 of them qualified scientists and engineers throughout the UK.[21] Nevertheless, in spite of its size and links with the parent company in Holland (where semiconductors are also mass-produced), Mullard has been forced into a defensive position, producing specialised devices for a limited market, and, like the indigenous British semiconductor companies, has been dogged with significant technical lag compared with the Japanese and American device manufacturers.

Amalgamations and mergers have been relatively few within the history of the company. Mullard purchased Associated Transistors in 1961, and in the following year formed Associated Semiconductor Manufacturers (ASM) in conjunction with GEC. This link with GEC was, however, broken in 1967. In that year, Mullard bought Pye Ltd., including their existing semiconductor plant at Newmarket.

An important factor distinguishing the Mullard company from other British semiconductor firms is that it is part of the Philips organisation. Based at Endhoven in the Netherlands, Philips is Europe's largest semiconductor company. It is vertically integrated in structure, manufacturing a wide range of electrical and electronic equipment, including television tubes and sets, semiconductors of a wide variety of types including a wide range of infra-red detectors for scientific, military and industrial applications, magnetic materials, and thermionic valves. It owns two companies in the United States, namely Siliconex and Amperex, and by 1980 possessed five major wafer fabrication plants. Each production centre concentrates on producing certain groups of devices in some particular laboratory, and consequently a multiplicity of device types are produced. The organisation operates on a world-wide scale; for instance, Signetics diffuses devices in the United States, assembles them in Korea, the Philippines and Thailand, also assembles military devices in the United States, and has a service centre in West Germany.[22] The international character of this company permits a wide range of flexibility in its policies, for example, in the early 1980s much discrete semiconductor and integrated circuit assembly was transferred from Europe to South East Asia, the French division RTC being particularly affected.[23] A current series of joint ventures involve Du Pont (optical discs), AT & T (Telecommunications) and Siemens (integrated circuits). In addition, Philips have obtained a foothold within the Japanese semiconductor market with a 35% holding in Matsushita Electronics.[24]

The international scale of the Philips operation brings into question any possibility of that company participating in a purely European venture aiming to establish an independent European electronic components industry, such as envisaged by some pro-European politicians. Yet without the participation of Philips,such an attempt would be doomed to certain failure. Certainly in any move of this nature, Philips would play a key role.

8.2.3 Ferranti Company

Ferranti Ltd. is a privately owned electronic component manufacturing company operating in the Manchester region with factories at Wythenshawe and also at Barrow-in-Furness. It is concerned with the manufacture of transistors, integrated circuits, optoelectronic devices, diodes, microwave valves and solid state microwave devices.[25] This company has been in the forefront of device development in Europe, as may be seen from Table 8.3.[26]

Certainly by any standards the development record of the Ferranti company is an impressive one. It is significant to note that Ferranti concentrated on silicon from the start, and there can be little doubt that this decision was to prove extremely advantageous at a later date when it became necessary to move into the field of planar device technology, as they possessed by this time considerable expertise in constructing devices using this material. Also, from the point of view of being able to get the new products into production, the Ferranti company has often been extremely successful. Yet, in spite of these advantages, commercial success has been somewhat limited. Golding, writing in 1972, states that 'A general survey would point to weakness in production and marketing' and goes on to suggest that the probable cause of these deficiencies is a shortage of cash arising directly from Ferranti's status as a private company and consequentially lacking the resources to fully exploit its technological ability.[27] He supports this statement by evidence that Ferranti expenditure in research and development at that time was lower in absolute terms than any comparable company.[28]

A disadvantage faced by Ferranti, in common with other European semiconductor manufacturers, is that because of the relative weakness of the home market, it must export the greater proportion of its products, and therefore compete directly with overseas semiconductor manufacturers. Writing in this context, the *Ferranti Review* stated that:

> 'The integrated circuit business is international and the UK sector accounts for only 5% of the total market. Europe accounts for 25% and the United States for 50%.'[29]

Table 8.3 Chronological development of devices

Date	Development
1953	Silicon chosen as the basic raw material for semiconductor production
1954	Low power silicon diodes now in production
1957	Silicon photovoltaic cells available commercially for the first time in Europe
1958	The first commercially available range of silicon mesa devices manufactured in Europe
1961	Silicon planar technology applied to volume production for the first time by a European company
1962	The first self-developed and manufactured range of integrated circuits (Micronor 1) D.T.L. Logic
1965	A range of D.T.L. integrated circuits produced which were claimed to be the fastest in the world, went into production
1967	Ferranti the first company in Europe to wholly develop and produce in volume a range of logic based on T.T.L. technology (7400 series)
1971	Development of uncommitted logic arrays now begun
1976	The first 16-bit microprocessor to be wholly developed, designed and manufactured in Europe

In order to improve its position in the US market, Ferranti acquired an American subsidiary, Interdesign, based in California, in December 1977, and this strategy appears to have been successful. By 1980, sales in the USA had more than quadrupled, and the total turnover in semiconductor products in that year amounted to about £28 million, over half this figure being accounted for by integrated circuit sales. By 1979, Ferranti was the largest and probably the only profitable British integrated circuit manufacturer.[30] By August 1981, the company held the largest single share (30%) of the uncommitted logic array market.[31]

Summarising the company's performance at that time, it would appear that, although quite strong in the field of research and development, weaknesses existed in the production area, and performance was inhibited by lack of a home market of sufficient size, and in addition, cash flow problems seemed to have caused difficulties. These problems however are not peculiar to Ferranti, but appear to conform to a general pattern observable within the European semiconductor industry.

8.2.4 The Plessey Company Ltd.
This firm began as a radio component manufacturer, being founded in 1925. However, during and after the Second World War it produced an increasing variety of electronic equipment, and has since been particularly active in the field of telecommunications.

Interest in the semiconductor field began relatively early. Representatives from Plessey (then the Automatic Telephone & Electric Company) attended the Bell technology symposium in April 1952, and research in the field of semiconductor technology began immediately, a programme on crystal growth being started in that year. This work resulted in the first silicon crystals being pulled from the melt in 1953. In 1957 a pilot plant was set up to meet the firm's growing need for semiconductor material.[32] From that year onwards, Plessey manufactured silicon alloy rectifiers at their Towcester factory, using a batch process modified from that developed by the American General Instrument Company.

Also in that year the Philco corporation's continuous process for manufacturing micro-alloy diffused high-frequency transistors was set up jointly with Plessey at the latter company's Swindon factory, under the title Semiconductors Ltd. This micro-alloy diffused (MADT) process was also used from 1961 onwards for making silicon transistors.[33] The venture was however unprofitable and soon rendered obsolete by the introduction of planar devices; nevertheless, production continued until 1967. Golding mentions that production yields at the Swindon factory were consistently lower than those achieved by the parent company in the United States and consequently prices quoted by Plessey were correspondingly higher.[34] He also quotes a market report which states that the company marketing activity was at that time 'probably weaker than that of any comparative maker.'

The Plessey company were the first British semiconductor manufacturer to become involved in the development of integrated circuits. A government contract was placed in April 1957 for the development of silicon integrated circuits on the lines indicated by G.W.A. Dummer of the Royal Radar Establishment, Malvern.[35] This development was carried out at the company's research laboratory at Caswell, Northamptonshire and was restricted to a very small team. Roberts states that:

'possible action at Caswell on the true silicon solid circuit was delayed by the departure of J.T. Kendall to join Texas Instruments, breaking the continuity of speculation on solid circuits.'[36]

The first research programme on 'silicon solid circuits' was submitted to the UK Ministry of Defence in November 1958.[37] However, integrated circuits did not go into limited production at Plessey until as late as 1965. Certainly, as a result of this delay, a great opportunity to enter the integrated circuit field at an early date had been lost. Dummer, commenting on this situation, attributes lack of success in this venture to:

'a failure of the UK military to perceive any operational requirements for ICs and an unwillingness within UK companies to invest their own money.'[38]

However, after 1967, Plessey ceased to manufacture discrete transistors and concentrated entirely on integrated circuits, producing both digital and linear types, but with an emphasis on linear circuitry, and ignoring standard TTL logic, then in greatest demand. Although probably the first company in the world to fabricate linear circuits, this venture was not really a success, and by 1977 the company were looking for a purchaser for their semiconductor plant.

Discussions took place at that time with the National Enterprise Board (NEB), GEC and GI Microelectronics, although with negative results.[39] Total turnover in Integrated circuit manufacture in 1978 was estimated at $18 million, somewhat less than that of the Ferranti company, then running at about $24 million.[40]

The Plessey company continued to manufacture integrated circuits, their greatest strength being in the field of emitter coupled logic (ECL). Until takeover, they operated from the following locations - Plessey Semiconductors, Plymouth, UK, Plessey Semiconductors Ltd., Swindon, and Plessey Three–Five Group at Towcester, where the Allan Clarke Research Centre is situated.

Plessey tended to produce devices for the military and industrial sector, rather than the commercial market and unlike Ferranti and Philips did not invest in overseas manufacturing or assembly plant. Substantial funding from government sources was made in the field of research and development and resulted in some success in this area.[41] However, basic problems appear to have existed in the vital areas of production and marketing over a considerable period and this may have been due, at least in part, to the company's orientation as a supplier of military equipment. In September 1989 Plessey finally lost a four year battle for independence, its semiconductor business being shared between GEC and Siemens on an equal basis. The experience of Plessey suggests that the problem of technical lag cannot be overcome merely by investment in research and development, but must be supported by a strong production and marketing sector, and that in this industry, the avoidance of delay in getting devices from the development stage into large-scale production is vitally important.

8.2.5 The General Electric Company (GEC)

The fortunes of this company in the semiconductor field have been extremely varied. The present company's roots can largely be traced back to two sources: one, British Thomson Houston Ltd. (BTH) which operated during the 1940s and 1950s, and manufactured point contact diodes of excellent quality, many being exported to the United States after 1945. The other source was the present company, GEC, which also carried out research, development and manufacture of point-contact diode detectors during the war at their Hirst Research Laboratories, situated at Wembley. At this time, BTH, together with Siemens Edison Swan (SES) and Metropolitan Vickers, were under the control of a holding company, namely Associated Electrical Industries Ltd. (AEI) whilst trading and competing as separate companies. Following the Second World War, GEC set up diode production at their radio factory at Coventry, where germanium point contact and alloyed junction transistors were manufactured. In the late 1950s, BTH were manufacturing high power devices at their manufacturing plant at Rugby, and also microwave and low power devices at their Lincoln factory. SES, GEC and Thorn were manufacturing in London, English Electric (which purchased Marconi in 1947) in Stafford, Marconi at Chelmsford, and Westinghouse at Chippenham (Wilts.).

Mergers commenced in the early 1960s, with BTH, SES and Metropolitan Vickers trading under the AEI name, with subsequent rationalisation of products. Other mergers were English Electric with Elliott Marconi and Westinghouse with Brush Electric. A working collaboration between English Electric and Westinghouse was formed to manufacture power transistors.

In the mid-1960s the semiconductor interests of BTH and SES were amalgamated and Thorn merged their interest also to form AEI–Thorn Semiconductors based at Lincoln. Marconi and Elliott also combined their interests to form Marconi–Elliott Microelectronics, based at Chelmsford and Glenrothes. Also at this time (1962), GEC and Mullard merged their semiconductor interests to form Associated Semiconductor Manufacturers (ASM) with production plants in London and Manchester. Work on integrated circuits commenced at Elliott–Automation in 1964 under a licensing agreement with Fairchild, and by 1967 Marconi–Elliott held 11% of the British integrated circuit market.

In 1968, Marconi formed an agreement with Ferranti to manufacture the latter company's Micronor II DTL logic. Large-scale product demand for this device did not however materialise. Elliott concentrated on production of the standard Fairchild 930 DTL series and they were in full production within the same year. TTL logic was also manufactured by Elliott at a slightly later date.

In 1967, GEC took over AEI and a year later merged with General Electric. The AEI–Thorn semiconductor association was broken, leaving AEI Semiconductors standing alone. As a result of these mergers, the following semiconductor plant operated under GEC management, AEI Semiconductors based at Lincoln, GEC Semiconductors at Wembley, Marconi Space & Defence at Portsmouth and the Hirst Research Centre at Wembley. There were still agreements with Mullard via ASM and with Westinghouse via English Electric at Stafford. However, these interests were eventually severed, that with Mullard in 1969, when the Hirst Research Centre reverted to GEC

Marconi Electronic Devices Ltd. was formed in 1980 from AEI Semiconductors to make power semiconductors and microwave components, GEC Semiconductors manufacturing custom integrated circuits and the MSDS Division, Portsmouth, hybrid circuits. An added acquisition, Circuit Technology Inc. of New York, also manufacture hybrid circuits.

Subsequent expansion of Marconi Electronic Devices Ltd. has included the acquisition of Tektronics of Swindon, who make telecommunication hybrid circuits, and Marconi Specialised Components at Billericay, manufacturing passive microwave components. Marconi Electronic Devices at Lincoln currently produce a variety of power devices, including high voltage and current switching transistors and thyristors.[42]

8.3 French semiconductor industry

The electronics industry in France has, in general, faced problems common to other European national semiconductor industries, namely, lack of an adequate home market and heavy foreign economic penetration, including fabrication and assembly plant operating within its shores. Like its counterparts in Britain and Germany, the French thermionic valve manufacturing industry largely took over the fabrication of transistors (new firms controlling only 10% of production in 1968)[43], thus conforming to the general European pattern.

From the early 1960s the American-owned firms IBM and Bull–GE dominated the French computer market.[44] (The French attempt to set up an indigenous computer industry collapsed when Bull was bought out by General Electric.) Commenting on this situation, Servan-Schreiber, writing in 1967, remarks that:

> 'what interested GE were the sales and maintenance outlets of those
> European firms, not their relatively weak technological potential.'[45]

These computer firms were in turn supplied to a large extent by American semiconductor plants which had been established on the shores of France. By 1967, no less than five US companies had transistor manufacturing facilities either operating, or just about to open, these being Texas Instruments (TI), Fairchild, ITT, Motorola and Transitron.[46] At that time, only two all-French semiconductor companies existed, namely, the Societé Industrielle de Liaisons Electriques (SILEC) and the Compagné Generale des Semiconducteurs (COSEM), a CSF subsidiary. In addition to these, Soveur Electronique SA was French controlled (although the American firm Corning Glass held a substantial minority interest). Two further firms were also engaged at this time in semiconductor manufacture, La Radio Technique SA, which was controlled by Philips and the Societé Europeanne des Semiconducteurs (SESCO), this latter firm being a joint venture between Thomson Houston and the General Electric Company.[47]

Work on integrated circuits started about 1960–61 at SESCO and units had reached the production stage by 1964. These were multi-chip circuits, the logic types diode-coupled transistor logic (DCTL), resistor-coupled transistor logic (RCTL) and diode–transistor logic (DTL) then being available on a fairly small scale. The first volume production of integrated circuits in France was started by Sovcor in 1968, the logic types then manufactured being DTL and transistor–transistor logic (TTL). At that date, most of the integrated circuits being constructed were for French military use.[48] Co-operative efforts were made at this time to strengthen the French semiconductor industry in the face of overseas competition. For example, COSEM and SILEC made an agreement in 1966 to co-operate on integrated circuit research and production. Also, at about this time, collaboration between COSEM and SESCO in semiconductor research was initiated and funded by the French government.[49]

However, in spite of French government support, by 1977 production by indigenous companies still amounted to less than 50% of the national market, foreign (mainly American) companies being responsible for the remaining total.[50] The value of the French market in integrated circuits in that year amounted to $130 million, about 20% of total European production. At this time, the government policy of encouraging mergers resulted in the formation of Thompson-CSF/SESCOSEM with the object of concentrating on the development of bipolar integrated circuits. To facilitate this object, a licensing agreement was made with General Instrument, National Semicon and Motorola in order to take advantage of American expertise in this field.[51] Also in that year, the French government produced a plan for the development of integrated circuits entitled 'Le plan circuits integres', the cost being in the region of $150 million. Regarding this event, Mackintosh writes that 'resulting from decisions which were made, the skeleton of a French microelectronics industry began to emerge by the end of 1978'.[52] Government support was extended in 1978 to include the construction of a new research centre situated at Grenoble to specialise in the development of integrated circuits. By 1982, advanced work was being carried out at this centre on MOS field-effect devices. Generally, government support policy regarding the electronics industry in France (as in Britain) has been to assist the larger manufacturing companies and to encourage amalgamation.[53]

In common with other European countries, the French semiconductor industry has invested in its American counterpart, in order to obtain the economic advantages of a production base within that country, in addition to advanced technical expertise. The major French investment to date occurred in 1979, when the largely French conglomerate Schlumberger acquired Fairchild.[54] Since that time, however, the fortunes of Fairchild have fallen into a relative decline, as can be seen from Table 6.3, and in October 1986, control of that company passed to the Japanese firm Fujitsu.

Most French semiconductor firms are centred in the vicinity of Paris, with a smaller number operating at Grenoble. La Radio Technique SA is situated at Caen, and Texas Instruments at Nice. In addition to these industrial research, development and production facilities, several French universities participate in semiconductor research and development.[55] Unlike the semiconductor industry in Britain and West Germany, there is therefore some evidence of the formation of skill clustering, although on a much smaller scale than in the United States. Further American companies now operating in France include ITT at Colmar, Westinghouse CDSW at le Mans and TRW Composants Elctroniques SA at Bordeaux.

The French government, perhaps more than its European counterparts, has attempted as a matter of policy to create conditions for the development of a nationally independent electronics industry, including the vital and interrelated areas of computing and semiconductor device production. So far, this policy appears to have had little success, and a significant factor cannot fail to be the relatively small scale of the operation. Putting aside questions of managerial and technical expertise, the sheer economy of scale and vast resources available to American industry have doomed to failure French attempts to establish an independent, nationally viable electronics industry. Certainly French resources could not have matched, for example, the investment by only one American company alone, IBM, of $5 billion during the early 1960s over a period of four years to launch the series 360 third generation computer.[56] Nevertheless,a modern semiconductor industry has been built up and, in addition to commercial sales, is able to supply at least a large proportion of French military integrated circuit requirements. Also, advanced research facilities have been established and are in operation.

8.4 Italian semiconductor industry

The history of the Italian semiconductor industry is largely that of one company, the Societa Generale Semiconduttori (SGS), which was founded in 1957 by Olivetti and Telettra, firms engaged in the business equipment and telecommunications areas, respectively. The original function of the company was to supply germanium diodes for the internal requirements of the parent companies.[57]

In 1961, Fairchild negotiated an agreement with the founding companies, receiving a one third share of the company in return for technical assistance in silicon planar technology.[58] This new organisation, renamed SGS–Fairchild, operated from Agate, near Milan, and subsequently expanded, setting up manufacturing subsidiaries in Britain, France, Germany and Sweden. (The British subsidiary was located at Falkirk, and full-scale production, largely on digital integrated circuits, commenced in 1968. This factory was however later closed in the early 1980s.) Dummer, writing in December 1964, mentions that at this time SGS–Fairchild was producing micro-logic circuits, although whether they were actually made in Italy at this time or imported from the United States is not clear.[59] Certainly, with the backing of Fairchild technology, SGS–Fairchild held a strong position at an early stage in integrated circuit production, holding 21% of the British market in 1967, second only to Texas Instruments who then held 25%.[60]

Golding suggests that the original Fairchild conception of SGS–Fairchild was that of a firm predominantly acting as a selling agency for American imports. This view is also held by Servan-Schrieber.[61] However, Fairchild were persuaded to set up production facilities in Europe, and furthermore, under pressure from Olivetti and Telettra, a research and development facility was established at Agate. The policy of both Telettra and Olivetti was to set up an independent research capability, and efforts on their part to expand the research and development facility at Agate led to further conflict with Fairchild. To quote Golding:

> 'when it became clear to Fairchild that it would not be able to acquire a majority interest in SGS–Fairchild, it initiated negotiations for the sale of the holding.'[62]

The sale took place in September 1968, resulting in a complete takeover by Olivetti, and the company then reverted to its original title, SGS. By that year, SGS had captured about 12% of the European semiconductor market and claimed leadership in silicon devices. Overseas expansion at this time included an assembly plant in Singapore, which began to produce low-cost discrete devices in 1969. A further factory is now in operation in Catania, Sicily.

Another firm operating in Italy at that time (but on a smaller scale) was the Siemens subsidiary ATES which was founded in 1957, and licensed to use RCA technology. When Olivetti decided to withdraw from SGS, the State telecommunications holding company bought out the latter company, merging it with ATES in 1971. This was done to prevent a takeover of SGS by Motorola.[63] The new company thus formed was titled SGS–ATES Mackintosh states that:

> 'by 1976 the company had established an advanced linear integrated circuit capability and product volume, complemented by advanced MOS production investment.'

In 1977, the Italian government passed an act entitled 'the Industrial Restructuring, Rationalisation and Development Law', providing $1.2 billion financial support to the electronic industry over a period of four years. The principal recipients were 'the electronics components sector, being SGS–ATES and Texas Instruments, providing the latter company established R & D facilities in Italy'.[64] (Texas Instruments Italia SpA currently operate from their location at Rieti.)

However, by the beginning of the 1980s the share of the European market held by SGS–ATES had declined owing to American competition and its response was to increase its operations within the United States by opening an extra number of sales offices in that country. This policy led to some measure of success, Mackintosh stating that sales to the US market doubled between 1980 and 1981.[65] In 1981, world sales amounted to $178 million, ranking SGS–ATES as the twentieth largest semiconductor manufacturer in the world, and the third largest in Europe, next in size to Philips and Siemens.

SGS–ATES overseas operations have been fairly widespread; for example, from 1961 onwards, manufacturing facilities were set up in France, Germany, Sweden and Britain, the latter being located at Falkirk in Scotland. (This plant, then manufacturing CMOS devices, was closed down in 1981.) Overseas assembly plants have since been at Singapore, in Malaysia, and at Malta. In addition to these units, the SGS Semiconductor Corporation also currently operates from Phoenix, Arizona.

Early in 1982, an agreement was signed between the Japanese firm Toshiba and SGS–ATES involving the transfer of CMOS technology from the former to the latter company, this being the first major agreement of the kind between a Japanese company and a European manufacturer.[66]

The varying fortunes of SGS–ATES illustrate the difficulties faced by a European firm attempting to compete in the American dominated international semiconductor market. In order to compete effectively, it is necessary to have access to the most recent technology and this can only be done by a policy of licensing agreements or mergers with the technically advanced American (or Japanese) semiconductor firms, or alternatively by developing separate research development and production facilities at such financial cost that it cannot fail to impose a considerable strain on the economy. The history of SGS–ATES shows that any attempt to establish an independent, balanced organisation including research and development, rather than merely acting as a fabrication and distributing agency will be strongly resisted, the non-national company providing the technical expertise responding by non-co-operation and withdrawal.

Italy, in common with all other Western European countries, does not possess the resources needed to establish an independent viable semiconductor industry, able to compete successfully in the international field. Its government-supported operation SGS–ATES has therefore been compelled to adopt a strategy of relying to a great extent upon foreign technical expertise.

8.5 The West German semiconductor industry

The West German semiconductor industry is now the strongest in Western Europe and is similar in structure to that of its other European counterparts, with the exception that it had had virtually no military market to sustain its activities. In 1968, the vertically integrated West German thermionic valve companies held 66% of the semiconductor market, a greater proportion than either Britain (32%) or France (42%), strongly conforming to the general pattern of the industry outside the United States.[67]

The electrical industry is dominated by two large firms, Siemens and AEG Telefunken, both vertically integrated companies with a wide range of interests in the electronics field. Work on integrated circuits began at Siemens in the early 1960s, and by 1964 that firm was producing small quantities of devices for internal company use. Also by this time Telefunken and Standard Electrik Lorenz (SEL), based at Nuremburg were producing large quantities of integrated circuits for digital applications,[68] and development work was proceeding at the Valvo plant in Hamburg, the latter firm being a Philips subsidiary.[69] The German approach to semiconductor device technology appears to have been one of caution at this time. For instance, Dummer writes

> 'In general it is the German practice to make full use of technologies already developed, as they feel that there is little point in making equipment obsolete until there is good reason to do so.'[70]

This approach may well have reflected the lack of available government contracts to stimulate the development of new devices.

As in other Western European countries, government support for the semiconductor industry has been made available. West German government assistance is supplied through the Ministry for Research and Technology of the Federal Republic. Their annual budget between 1971 and 1978 showed a gradual increase from the 1971 figure of 10·4 million D. to 97·4 million D. in 1978, these sums including financial aid in respect of both discrete components and integrated circuits.[71]

By 1980, Siemens was beginning to emerge as a significant supplier of components to the United States and also by this time AEG Telefunken had become by far the largest manufacturer of discrete components in the European Economic Community.[72] In Germany, as in all other European countries, the integrated circuit market has been dominated by the United States, but in spite of this situation, the West German semiconductor industry, and Siemens in particular, was investing heavily in the American semiconductor industry during the late 1970s.[73] By 1982, Siemens controlled three American firms outright, namely, Dickinson Associates, FMC, and Litronix, held a 20% share in Threshold Technology, and operated overseas assembly plants in Singapore and Malaysia.[74] At about that time it closed down its component manufacturing facilities in Mauritius and Arizona.[75]

Current American investment in the German semiconductor industry includes Texas Instruments GmbH at Freisling, where integrated circuits are currently being manufactured,[76] Motorola GmbH Semiconductor Products which operates from Unterfahring and Intermetall (ITT) GmbH who are located at Freiburg.[77] In addition to American economic penetration, Japanese investment has also taken place, and the following production facilities are currently in operation: Hitachi Semiconductors (Europe) GmbH, situated at Landshut, Bavaria, and Toshiba Semiconductor GmbH, at Braunschweig.

What is evident from above is the international nature of the German semiconductor industry, which, although the strongest in Western Europe, has undergone significant economic penetration from overseas. Regarding the position of Siemens, the largest German semiconductor manufacturer, the Mackintosh profile of the world-wide semiconductor industry, published in 1982, comments that:

> 'in common with other European electronics companies, Siemens has lagged behind American and Japanese firms in innovation and marketing.'[78]

It therefore appears that problems of overseas economic penetration, lack of investment, deficiencies in marketing and technical lag are all present to some extent within the German semiconductor industry, as also in the case of its other European counterparts.

8.6 Soviet and East European semiconductor industry

The semiconductor industry within the Soviet Union has grown up under somewhat different conditions to that existing in the non-Communist world, with its international market economy. A matter of importance affecting its growth has been certain demands made upon the internal industrial economy which have arisen increasingly during the last two decades. In this context, Amann writes that:

> 'Technological inertia has come to be seen, both among Western analysts and in the USSR itself, as perhaps the crucial constraint on future Soviet economic development.'[79]

Certainly, surrounded by a politically hostile environment, the Soviet Union has been forced to respond to industrial developments outside its frontiers, and attempt to achieve technical parity in a wide range of fields, including semiconductor device technology.

In view of this problem of technical inertia, the prospect of automating the Soviet manufacturing industry and thereby overcoming existing shortcomings by raising productivity, whilst at the same time improving reliability and quality through the introduction of advanced electronic equipment and processes, would be highly desirable. In this context, Snell writes that a significant factor in the introduction of microprocessors into the Soviet economy was the possibility of providing a solution to the inefficient operation of production enterprises.[80] Also, in addition to commercial and industrial applications, the prospect of using advanced electronic equipment for military purposes might be viewed by the Soviet authorities as essential in view of the existing world political situation. As in the United States, military requirements must continue to play an important role in stimulating semiconductor device development. Unlike the United States however, the economic need for the industry to expand its operations into a growing overseas market does not exist, nor for political reasons is the Soviet industry likely to assemble devices in Third World countries where wage rates are low. Compared with the United States, therefore, the semiconductor industry in Soviet Russia is economically disadvantaged, in terms of unit cost. This factor can only act to discourage the export of semiconductors from the Eastern bloc to the West, although a significant number of semiconductors have fairly recently been exported to the United States, and are now currently being offered for sale in the United Kingdom.[81] Since the history of the Soviet semiconductor industry is largely a response to developments outside its frontiers, it is not surprising that a technical lag should exist. This lag has variously been estimated as between two and ten years although it must vary greatly from one sector of the industry to another.[82] In the case of the microprocessor, it has been quoted as being in the order of about four years.[83]

Some evidence exists that transistors were being manufactured within the Soviet Union fairly soon after their invention in the United States. G.P. Kazanski of the Collegium Radio Technical Ministry in Moscow, during a visit to America in 1955, stated that transistors were currently replacing vacuum tubes to about the same extent as within American industry and that they were being largely used at this time in the construction of measuring instruments and computers. Although no details of device manufacture were supplied, Kazanski also mentioned that these computers were mainly used for statistical work.[84] However, despite this early start, advances in solid state technology do not appear to have kept pace with developments outside the Eastern bloc.

An important factor affecting the growth of an indigenous Soviet semiconductor industry cannot fail to have been the trade embargo, imposed by the United States and its allies, in an effort to deny the Soviet Union access to advanced Western technology. To what extent development of the microelectronics industry has been assisted by acquisition of technological knowledge from the West is impossible to evaluate,[85] but lack of easy access to this knowledge may well have stimulated the growth of the indigenous industry. Certainly all the main logic families produced in the West are currently being manufactured in the Soviet Union, namely, Shottky TTL, I^2L, nMOS, pMOS and CMOS. Much information regarding computer manufacture has been obtained from Western sources, for example, microprocessors such as the K 580 and K 589 series, based on the American Intel Corporation MCS 80 and 3000 series respectively, are in existence.[86] However, many other types of microprocessor exist which do not appear to have a Western counterpart[87] including a 16 bit single-chip microcomputer,[88] also currently in use. It now appears that in addition to building up a viable device manufacturing industry, a broadly based computer industry has come into existence.

Development of the semiconductor industry in Eastern Europe is coordinated through the Council for Mutual Economic Assistance (CMEA). Participating countries are the Soviet Union, the GDR, Poland, Czechoslovakia, Hungary, Romania and Bulgaria. This organisation was set up in 1982 and covers the development, production and applications of hardware and software. The organisation aims to avoid wasteful repetition of work and to meet the members' equipment needs.[89] One result of this policy was the transference of a semiconductor plant manufacturing LSI and VLSI devices from the Soviet Union to Hungary in 1984. Two production lines were installed, producing about 20 million integrated circuits and 80 million other microelectronic components per annum.[90] In addition to the diffusion of technology, the organisation possesses an economic dimension; for example, CMEA policy is to provide reduced customs tariffs to parts imported from other member countries.[91] One result of CMEA strategy is that computers of various types are being currently manufactured in the GDR, Czechoslovakia, Hungary and Bulgaria.[92]

8.7 Summary

This review of the major European national States shows that, compared with the situation in the United States and Japan, the developing Western European semiconductor industry has suffered significant disadvantages; namely, a general lack of cohesion, each government pursuing a separate strategy with no attempt being made to work to a co-ordinated plan. Financial institutions have, in varying degrees, proved inadequate in providing sufficient financial support to enable a healthy industry to develop. The general weakness of the Western European semiconductor industry has been accentuated by strong foreign penetration of the home market, and this situation has been assisted by lack of a common response from Western European governments. With research and development facilities often underfunded and with relatively small investment within the important area of production development, a significant technical lag has remained a constant factor throughout the history of the industry.

Like its Western European counterpart, the semiconductor industry in Eastern Europe is relatively weak when compared with that of the United States and Japan, although for different reasons. Confronted with a strategic trade embargo, and denied the dubious benefits of foreign economic penetration, a consistent technical lag has also been present, in spite of an early beginning in device manufacture. Owing to lack of information regarding funding of the Eastern European industry, it is impossible to assess the effect this factor may have had upon its development.

8.8 References

1 *Electronics Weekly* 28 January 1987, p. 2
2 FINAN, W.: 'The international transfer of semiconductor Technology through US based firms' (National Bureau of Economic Research, NY, 1975) p. 94
3 GEE, G.: 'World trends in semiconductor developments and production', *British Communications and Electronics*, 6th June 1959, pp. 450–51
4 'Microelectronics into the 80s' (Mackintosh Publications, 1979) p. 24
5 GOLDING A.M.: 'The semiconductor industry in Britain and the United States: a case study in innovation, growth and diffusion of technology'. D. Phil Thesis, University of Sussex, 1971; See table entitled 'Firm shares of the US semiconductor market', p. 179
6 TILTON, J.: 'International diffusion of technology; the case for semiconductors' (Brookings Institute, Washington DC, 1971) p. 34
7 Survey by FINAN, W.: 'The international transfer of semiconductor technology through US based firms' (National Bureau of Economic Research, NY, 1975) pp. 50–52
8 OECD Gaps in technology series: 'Electronic components' (OECD, Paris, 1968) p. 82

9 BRAUN, E., and MACDONALD, S.: 'Revolution in miniature' (CUP 2nd Edn., 1982) p. 163; also conversations with G.W.A. Dummer, T.R.E. (Malvern) and T.G. Brown (Mullard Ltd., Southampton). See also A.M. Golding, *op.cit.*, p. 379.
10 DONALDSON, P.: 'Guide to the British Economy' (Penguin Books), p. 47
11 BRAUN, E., and MACDONALD, S.: *op.cit.*, p. 159
12 GOLDING, A.M.: *op.cit.*, p. 348
13 'Microelectronics into the 80s' (Mackintosh Publications, 1979) p. 34
14 GOLDING, A.M.: *op.cit.*, p. 261
15 GOLDING, A.M.: *Ibid.*, p. 265
16 'Microelectronics into the 80s' p. 35
17 OECD: 'Gaps in technology', *Ibid.* p. 75
18 *Ibid.*, p. 45
19 GOLDING, A.M.: *op.cit.*, p. 179
20 *Ibid.*
21 Data supplied by Mullard Ltd.
22 'Profile of world-wide semiconductor industry' (Mackintosh Publications, 1982) p. 41
23 *Ibid.*
24 *Electronics Times*, 2nd October 1986
25 *Ferranti News*, June 1976
26 *Ibid.*
27 GOLDING, A.M.: *op.cit.*, p. 212
28 *Ibid.*, p. 212
29 *Ferranti Review*, Aug. 1980
30 *Ferranti News*, Aug. 1981
31 'Microelectronics into the 80s', *Ibid.* p. 24
32 WILSON, B.L.H.: '25 Years in silicon technology', *Systems Technology*, March 1951, (33)
33 WILSON, B.L.H.: *Ibid.*
34 GOLDING, A.M.: *op.cit.*, p. 216
35 DUMMER, G.W.A.: 'Integrated Electronics' *Electronics & Power*, March 1967, p. 73
36 ROBERTS, D.H.: 'Silicon integrated circuits' *Electronics & Power*, April 1984, p. 282
37 ROBERTS, D.H.: *Ibid.*
38 DETTMER, R.: 'Prophet of the integrated circuit' *Electronics & Power*, April 1984, p. 287
39 'Microelectronics into the 80s', p. 24
40 *Ibid.*
41 GOLDING, A.M.: *op.cit.*, p. 215
42 EVELEIGH, R.D., Publicity Officer, Marconi Electronic Devices, Lincoln (Private communication)
43 BRAUN, E., and MACDONALD, S.: *op.cit.*, p. 160
44 *Electronics*, 26 December 1966, p. 90
45 SERVAN-SCHREIBER, J.J.: 'The American Challenge' (Penguin) p. 113
46 *Electronics*, 26 December 1966, p. 90
47 DUMMER, G.W.A.: *Proc. IEEE*, December 1964, p. 1422
48 *Electronics*, *Ibid.*
49 *Ibid.*
50 'Microelectronics into the 80s' pp. 7–10
51 *Ibid.*
52 *Electronics*, 15 December 1982
53 BRAUN, E. and MACDONALD, S.: *op.cit.*, p. 164
54 US International Trade Commission: 'Competitive factors influencing world trade in integrated circuits'. Washington DC, 1979, p. 106
55 DUMMER, G.W.A.: 'Integrated electronics in Europe', *Proc. IEEE*, Dec. 1974, p. 1419
56 SERVAN-SCHREIBER, J.J.: *op.cit.*, p. 113
57 GOLDING, A.M.: *op.cit.*, p. 225
58 *Ibid.* p. 225
59 DUMMER, G.W.A.: *Proc. IEEE*, Dec. 1964, p. 1422
60 GOLDING, A.M.: *op.cit.*, p. 156
61 SERVAN-SCHREIBER, J.J.: *op.cit.*, p. 113
62 GOLDING, *op.cit.*, p. 227

63 'Profile of world-wide semiconductor industry' (Mackintosh, 1982), p. 45
64 'Microelectronics into the 80s' (Mackintosh, 1979), p. 14
65 'Profile of world-wide semiconductor industry', p. 46
66 *Ibid.*, p. 46
67 TILTON, J.: 'International diffusion of technology', *Ibid.*, pp. 115, 144
68 An offshoot of the American Company International Telephone and Telegraph (ITT)
69 DUMMER, G.W.A.: 'Integrated electronics in Europe', *Proc. IEEE*, Dec. 1964, pp. 1413–22
70 *Ibid.*
71 'Microelectronics into the 80s' (Mackintosh, 1979) Table, p. 51
72 *Ibid.*
73 BRAUN, E. and MACDONALD, S.: *op.cit.*, p. 167
74 'Profile of world-wide semiconductor industry'. Table, p. 108, also see p. 73
75 *Ibid.*, p. 48
76 Conversation with J. Brown (Texas Instruments, Bedford)
77 Listed in 'International semiconductor technology survey of production facilities' *in* 'Profile of world-wide semiconductor industry' (Mackintosh, 1986)
78 'Profile of world-wide semiconductor industry' p. 49
79 AMANN, R.: 'Setting the Scene' *in* 'Technical progress and Soviet economic development' (Blackwell, 1986) p. 5
80 SNELL, P.: 'Soviet microprocessors and microcomputers', *op.cit.*, p. 53
81 *Elorg*, 1986, 1 11, p. 22
82 WOHL, R.: 'Soviet research and development', *Defense Science and Electronics*, Sept. 1983, p. 11
83 SNELL, P.: Personal communication
84 'Industrial Report', *Electronics*, Jan. 1956, p. 22
85 SNELL, P.: *op.cit.*, p. 62
86 SNELL, P.: *Ibid.*, Table, p. 73)
87 SNELL, P.: *Ibid.*, p. 54
88 'Annual update of Microprocessors', *EDN*, 5 Nov. 1980, p. 180
89 SNELL, P.: *Ibid.*, p. 61
90 'Hungary: Market Report' (British Embassy, undated)
91 *Ibid.*
92 *Elorg, op.cit.*, p. 4

Chapter 9
Conclusions

The object of this chapter, as stated in the Introduction, is to bring together the material and ideas present in the preceding chapters, and to portray the result in the form of a coherent set of conclusions.

It is also of importance, however, to draw attention to certain characteristics of the semiconductor industry which set it apart from previous technical developments. Starting from the invention of the point-contact transistor little more than thirty years ago, the subsequent scale of growth of the industry has been unprecedented. A fundamental aspect of this event is that its consequences have been of a pervasive nature, leaving hardly any aspect of modern society unaffected. Certainly, the immense increases in power and reliability of computers, together with their ever decreasing size, could only have been achieved through the continual improvement of solid state devices. These technical advances in computing have led to widespread applications in such fields as civil and military communications, banking and office systems, machine tool automation, medical technology and domestic labour-saving equipment. Also, previously unforseen markets such as digital watches, hand-held computers, transistor radios and computer games have developed directly as a result of constant advances in semiconductor device technology. A far-reaching effect of the emergence of a semiconductor based computer industry has been the ability to receive, store, analyse and disseminate information on a vastly increased scale. With these achievements in mind, together with a realisation of the profound social consequences already being felt by society, an awareness of the importance of the rapidly developing semiconductor industry can hardly be too greatly stressed. It follows therefore that the need to understand and debate the problems being raised by these events has become a matter of urgent importance. The present book, in outlining the growth and development of the semiconductor industry on a world-wide scale, is intended as a contribution to that debate.

The following sections of this chapter consider two different aspects of the work separately. Firstly, a discussion of the growth of the semiconductor manufacturing industry and its relationship to science and technology; and secondly, a consideration of economic factors, operating within a social and political framework, which have contributed to the manner in which the industry has developed within various nation states, and ending with a brief speculation regarding future possible developments.

9.1 Growth of the semiconductor industry and its relationship with science and technology

The origin of the semiconductor industry has been traced to its beginnings in nineteenth century experimental science, and mention has been made that as early as 1883 selenium rectifiers capable of rectifying low frequency alternating currents had been fabricated on an experimental basis.[1] However, the first commercially important use of semiconductors dates from around 1904 when it was first discovered that these materials could act as detectors of high frequency currents.[2] Interest in radio communication at that time was intense and the 'cat's whisker' was certainly superior to other methods of detection then being used (for example the Branley–Lodge coherer) and quickly supplanted them. As a consequence of the success of this device, much experimental work took place in investigating the properties of various semiconductor materials. Although an adequate theory explaining the behaviour of rectification did not emerge at that time, the work of G.W. Pearce convincingly showed that the rectification effect was electrical rather than thermal.[3] Hence, even in the early stages of work in the field of semiconductors, a situation arose in which the success of the semiconductor detector in an area of current technological prestige stimulated fundamental work on the properties of semiconductor materials, leading in turn to an advance in theoretical knowledge and serving as an illustration of what Gibbons and Johnson describe as a symbiotic relationship between science and technology.[4]

Unfortunately the point-contact diode did not possess the ability to amplify an electrical signal, and when the triode valve was shown to have this property by Armstrong and Lieben (about 1911), the mainstream of attention focused on the properties of electron emission in a vacuum. This trend was reinforced by the demands of the First World War, during which time the vacuum tube industry was established in the countries of the main contestants (this industry utilised and developed mass-production techniques at this time which were to be of great value in the production of semiconductor diodes and transistors several decades later). Following the First World War, the broadcasting industry grew rapidly and thermionic valve production flourished. Under these conditions there was little incentive to attempt to improve the performance of crystal detectors and both theoretical and experimental work languished. However, some progress was made during this period in the field of diode rectification rather than signal detection. Although selenium rectifiers capable of rectifying low frequency alternating currents had been made as early as 1883–86 by C.E. Fitts[5], development of selenium rectifiers on a commercial scale did not occur until the late 1920s[6] following the introduction of the copper oxide rectifier by Grondahl and Geiger in 1927.[7] The commercial success of both rectifier types stimulated further work in this area,[8] contributions being made by Mott[9], Schottky[10] and Davydov[11] on the theory of rectification, including metal-to-semiconductor contacts.

A fundamental reason why relatively slow progress was made in solid-state theoretical research during the inter-war period was that, in order to investigate adequately physical phenomena in semiconductors, it was necessary to work with extremely pure materials and at this time there was no known means of producing them. Smith mentions that much of the work carried out on the metallic oxides and sulphides using compressed powders and evaporated or chemically deposited films frequently gave misleading results.[12] Under these conditions it is not surprising that progress in developing a theory to explain the physical behaviour of semiconductors was slow.

An important conceptual breakthrough came as a result of work carried out in the field of quantum mechanics in the late 1920s, the first application of this approach to the treatment of electrons in metals being due to A. Sommerfeld in 1928.[13] In this context, Smith writes:

> 'It was not till after this theory was available that a satisfactory definition of a semiconductor could be given.'[14]

The theory referred to by Smith was that due to A.H. Wilson and published in 1931, explaining the motion of electrons in crystalline solids.[15] Considerable uncertainty still existed however, Wilson's original papers expressing a doubt as to whether silicon was a semiconductor or a metal.[16]

It is therefore significant that progress in semiconductor research during the inter-war period was frustrated in two quite different ways: firstly, by the technical inability to produce sufficiently pure semiconductor material to enable consistent experimental results to be obtained, and secondly by lack of an adequate theoretical model to explain the behaviour of charge carriers in semiconductor material.

As we have seen in Chapter 3, the need for an effective replacement for the thermionic valve as a detector of high-frequency signals in the field of radar during the Second World War stimulated an increased effort to produce point-contact diodes with improved characteristics, and as many as 30 or 40 firms were then engaged in work on semiconductors in the United States alone. Part of this effort was a great increase in fundamental research on both silicon and germanium, the most fruitful work in this respect being carried out at Bell Telephone Laboratories and at Purdue University. The results of this research at Purdue on germanium (from which material the first transistors were made) were of particular value. Having a lower melting point than silicon, it was easier to purify, thereby facilitating experimental studies. The fact that only extremely small amounts of impurities vastly alter the properties of a semiconductor is of particular relevance, since it illustrates an important feature of semiconductor research, namely, the close dependence on the technology of material preparation. Indeed, in this field, techniques of material preparation have been vital in producing the means of verification of theoretical concepts.

The development of techniques of material preparation appears to have been greatly dependent upon the urgency of technological need. As an illustration of this, it is significant that the critical process of zone refining (discussed in Chapter 5) described in a paper by P. Kapitza in 1928[17] was ignored until, as W. Pfann writes 'the need arose for germanium of uniform purity in semiconductors.'[18]

This led Pfann (in the early 1950s) to consider the method of zone levelling–a technique he had conceived in 1939, but not pursued at that time. Not only did this approach enable greatly superior devices to be made on a mass-production basis, but the resulting existence of large crystals of uniform purity greatly facilitated subsequent theoretical work on semiconductor materials. This event furnishes an illustration of a technical process which would, with hindsight, have been invaluable in assisting theoretical studies of semiconductor materials during the inter-war years, but instead was neglected until a practical need arose (at a much later date) shortly after the invention of the transistor, for germanium of uniform purity.

Thus in the early history of semiconductor development a picture emerges, not of pure theoretical studies leading to the discovery of transistor action which was then exploited by technologists, but what appears to be a much more complicated and subtle process in which scientific and technological progress was interrelated, each assisting the other in turn. The main stimulus was, however, the need to solve problems raised by the failure of existing technology to satisfy a particular need; for example, the amplification of electrical signals, which resulted in the development of the thermionic valve, and again in its turn, the inability of the thermionic valve to function as a detector of very high frequency signals. This latter event led to renewed work in the semiconductor field which assisted by the recent theoretical developments in quantum mechanics, was able to make significant progress. Concerning this later effort, the overriding stimulus during the wartime period was military demand. Lavish government funding, particularly in the United States, enabled large goal-directed research teams to be set up, whilst at the same time emphasis was placed on producing new devices under conditions of extreme urgency. This resulted in the establishment, again particularly within the United States, of a large scientific–industrial complex, oriented towards military requirements.

The establishment of research teams was particularly important in the case of semiconductor device technology, since work in this field is of a multi-disciplinary nature, encompassing physics, metallurgy, chemistry and electronic engineering, and therefore crossing traditional academic boundaries. The situation by the end of the Second World War was that there now existed teams of scientists experienced in the field of semiconductor diode technology, well acquainted with the theoretical concepts of quantum mechanics as applied to semiconductors and experienced in working with colleagues trained in allied disciplines. It is significant that the invention of the point-contact transistor in 1947 was a product of one of these teams, engaged in goal-oriented research.

Although the existence of these trained teams of scientists specialising in semiconductor technology was a key factor in the invention of the transistor, it is unlikely that, without the insight provided by the theoretical concepts arising from quantum physics, significant progress could have been made towards the discovery of transistor action. In this context, it is notable that the internal Bell Laboratories document permitting authorisation for work, published in the summer of 1945, and which was to lead to the invention of the point-contact transistor, stated:

'The quantum physics approach to the structure of matter has brought about greatly increased understanding of solid-state phenomena. The modern conception of the constitution of solids that has resulted indicates that there are great possibilities of producing new and useful properties by finding physical and chemical methods of controlling the arrangement and behaviour of atoms and electrons which compose solids.'[19]

Bardeen and Brattain were, to quote Sparkes:

'members of a small group of physicists working under the direction of W. Shockley in fundamental research on the solid state of matter.'[20]

and their discovery of the transistor principle, according to Shive, took place:

'in the course of theoretical and experimental studies on the possibility of controlling the resistances of thin semiconductor layers by applications of electric fields strong enough to penetrate their surfaces.'[21]

Furthermore, according to Braun and Macdonald:

'the discovery had been made during an investigation of surface properties and was not the result of an exhaustive programme to use semiconductors to make a transistor.'[22]

These statements seem to suggest that the invention of the transistor arose as a result of the ability to recognise the potential importance of particular phenomena observed during the investigation of surface states in semiconductor material. Certainly, the practical applications of the observed phenomena (transistor action) were quickly realised and an amplifier was immediately constructed (27th December 1947).[23] Moreover, evidence exists of the realisation by the research group involved of the possibility of a semiconductor amplifier being produced as a result of this work.[24] Because of this realisation, perhaps the previously used term 'goal oriented research' best describes the conditions under which the transistor emerged.

However, at this stage much theoretical work remained to be done in order to produce an adequate theory of transistor action. In this context, R.A. Smith writes that:

'the transistor effect itself has proved to be a powerful tool for investigating the fundamental properties of these (semiconducting) substances so that the interaction between pure and applied research has been most fruitful.'[25]

Although the work leading up to the invention of the point-contact transistor took the form of goal–orientated research using existing solid-state theory and was firmly based on the twentieth-century concepts of quantum mechanics, the subsequent development of semiconductor devices has been overwhelmingly due to advances in the technology of device fabrication based on the junction transistor. Although this device, invented by Shockley in 1952, took longer to develop, it could be manufactured with stable characteristics. Furthermore, unlike the earlier point-contact transistor, its operation rested on a sound theoretical basis, enabling device behaviour to be predicted. This ability to predict the behaviour of device characteristics and successfully modify them was indeed the key to the subsequent development of the semiconductor industry. Consequently, the future of semiconductor manufacture was to be built on a theoretical understanding of semiconductor junctions and surfaces.

In the early stages of device development up to the invention of the planar transistor, several quite different approaches (although all based on the junction rather than the point-contact principle) were pursued with varying success, as discussed in Chapter 4. A common feature of the work of this stage was the inability of device manufacturers to overcome the undesirable effects of surface leakage current. The theory of surface states, highly complex in itself, could only furnish guidelines within which device engineers could operate. It was within this environment that much development work was carried out by device engineers sometimes with only a fairly basic knowledge of semiconductor theory but whose work nevertheless often achieved considerable success in improving electrical device parameters and production yields. Much of this knowledge was of an empirical nature and was greatly prized and often jealously guarded when successful. It was the ability of these device engineers, operating at this level, which was a determining factor in the commercial success of the product.

The invention of the planar transistor and the subsequent development of the integrated circuit took place almost entirely within the semiconductor manufacturing industry. Planar technology not only solved the problem of surface leakage, with consequent improvements both in electrical device parameters and reliability, but the mastery of this approach led directly to the ability to produce integrated circuits in ever increasing complexity. These improvements did not involve any basically new physical concepts, but relied upon the development of scientifically based industrial technology. The situation may be conveniently summed up in the words of Braun and Macdonald:

> 'The integrated circuit was a commercial innovation developed by scientists working in a technological society.'[26]

Therefore, unlike the situation within the years immediately preceding the invention of the junction transistor, subsequent advances have been mainly in the nature of 'technology building on technology', to quote again the words of Gibbons and Johnson—in this case, however, a technology of a highly advanced, scientifically-directed nature.

9.2 Emergence of the semiconductor industry on an international scale and its subsequent development

Although a range of photocells and small current rectifiers was being manufactured both in Europe and America during the 1930s, the establishment of semiconductor manufacture on an industrial scale really dates from the beginning of the Second World War. At that time, device manufacturing facilities were set up in the United States, the United Kingdom and Germany, in order to produce semiconductor detectors for use in the field of radar and also to manufacture photocells, which were mainly used as infra-red detectors.

The defeat of Germany effectively brought the industry to an end in that country and immediately following the cessation of hostilities, the demand for these devices in both the United States and Britain fell considerably. Nevertheless, one result of the war was that, owing to the sheer scale of American effort towards the end of that conflict, the centre of gravity of the electronics industry, including semiconductor research, development and production, shifted from Europe to North America. From this time onwards, until the mid 1980s, the US semiconductor industry was to dominate in this field on a world-wide scale, and the principle reasons for this situation are worth analysing.

Resulting from the invention of the transistor at Bell Laboratories in 1947, the American electronics industry was in a unique position to develop and market this device. Several important factors were to assist the Americans in this undertaking. Firstly, the defeat of Germany and Japan ensured that, at least temporarily, these technically advanced countries offered little competition. The sheer size of the American electronic device industry, reinforced by unparalleled numbers of trained scientists, engineers and technicians, gave a large advantage in economy of scale. Also, the American economy was relatively much stronger than that of Europe, capital resources were far greater, and much more capital was available for investment in new enterprises. Research and development facilities were present on a much larger scale than elsewhere and lavish government funding in the form of both grants to industry and defence contracts were also available on a similar scale. In addition to these advantages, a large potential home market existed, enjoying a much higher standard of living than in Europe and Japan, and consumer orientated to a far greater degree than elsewhere.

Most American semiconductor manufacture during the early years of production was centred on the home market. For example, as late as 1959, only just over 2% (by value) of total factory production was being exported.[27] In that year however, 45.5% of US semiconductor production (by value) was designated for military use.[28] During this early phase of semiconductor development military investment increased markedly, stimulated by the Korean war. This situation was to continue until 1960, when military purchases reached a peak of 48% of US semiconductor production (by value).[29] During this time, the electronics industry within the United States had undergone a vigorous expansion and this early growth of the semiconductor manufacturing industry should be seen as part of this trend (see Fig. 4.3).

A feature of the American semiconductor industry has been its sensitivity to the possibility of any threat from foreign imports. As an illustration of this attitude, the Japanese challenge in the cheap germanium market amounted in 1959 to only about 0·4% (by value) of the total US home market.[30] Nevertheless, the industry at this time was seriously worried by the volume of cheap Japanese germanium imports, the American Electronics Industries Association (EIA) complaining at that time to their government on the grounds of 'a present or foreseeable threat to our nation of security'.[31]

A significant factor in maintaining the American technical lead was that, owing to the military orientation of their industry, expertise in the field of silicon technology was relatively more advanced than in other countries. This was to be of particular importance in assisting the development of the silicon planar transistor, from which the integrated circuit was soon to be developed. It is important to remember that these developments did not take place in isolation, but in close relationship with the growing computer industry, largely in order to satisfy military demands. The need for lightweight, reliable systems was of particular importance in the field of rocketry and this requirement greatly stimulated the trend towards microminiaturisation. This trend was particularly significant in the years immediately before the invention of the integrated circuit, which may be seen as a response to an urgent need. For example, consider the trend from aircraft towards missile systems between 1955 and 1960. In 1955, aircraft represented 36·3% of total procurement and missile systems 5·4%. However, by 1960, aircraft represented 22·5% of total procurement and missile systems 27·4%. The importance of the integrated circuit in meeting this need can be seen from Fig. 6.3, indicating the percentage production of integrated circuits designated for military use. It is evident that, in the early years of production, almost all of these devices were purchased by the government. The integrated circuit industry, having received this initial boost, was soon providing these devices at vastly reduced prices, as evidenced by the data illustrated in Fig. 6.4. Sales of products utilising integrated circuits in both the industrial and commercial field grew rapidly, as new applications, such as hand-held computers were developed and found a ready market.

An important trend which developed from about 1960 onwards was the progressively increasing size of the semiconductor market outside the United States. Fig. 9.1 illustrates that between 1960 and 1978 the American share of world semiconductor consumption had fallen from 85% to 38%. Nevertheless, Fig. 7.3 shows that during this time the corresponding American share of the world semiconductor market remained fairly constant at about 60%. It is significant that American semiconductor operations overseas grew substantially over this period, both in terms of direct exports from the home base and as a result of setting up assembly and fabrication facilities in both Europe and Asia. This success is certainly evidence of the strength of the US semiconductor industry during this period compared with that of its overseas rivals and also of its capacity to adapt more quickly to changing world conditions.

The only significant challenge to American domination of the semiconductor market has come from Japan. The growth of the Japanese semiconductor industry, unlike that of the United States, has taken place under conditions of planned strategy, organised and directed by the Ministry of International Trade and Industry (MITI). The present success of the Japanese approach has been achieved by long-term loans to selected firms in industry, these firms being large vertically integrated manufacturers of electronic products, heavily financed, and assisted by government research and development facilities. The strategy has been to concentrate on selected areas of the industry such as very large scale integration (VLSI) in an effort to catch up and overtake the Americans within the chosen area. Having achieved this aim, the advantages of being technical leader then accrue. During the present decade, the American semiconductor industry has found it progressively more difficult to raise capital, and the required return on capital investment is well above that demanded by the Japanese banks, putting the American effort at a further disadvantage. Owing to the availability of long-term loans from Japanese banks, a decided advantage of their industry has been its ability to sustain financial losses over a

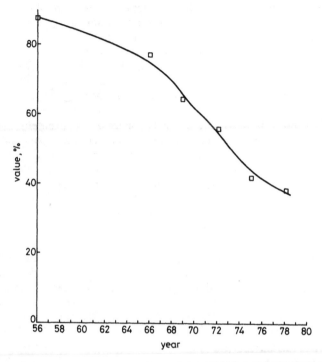

Fig. 9.1 Estimated world semiconductor consumption by USA versus the rest of the world (Source: Dataquest, 30 May 1979; and FINAN, W.: 'International transfer of technology through US based firms'. Nat Bureau of Economic Research, NY, 1985, p. 94

lengthy period. This has been of particular importance in view of the nature of the industry, which is prone to marked product-cycle variations.

From Fig. 7.3 it can be seen that the Japanese share of the world semiconductor market has increased since the beginning of the 1980s whilst that of the United States semiconductor industry has declined by a roughly similar amount. Furthermore, Japanese economic penetration of the American semi-conductor market is currently running at just under 20% of the market (by value) compared with a 10% penetration of the Japanese home market by the United States. Part of the Japanese strategy has been to resist penetration of the home market by discouraging foreign subsidiaries and also by preventing them acquiring controlling interests in Japanese firms. In addition, governmental measures have included the monitoring of licensing agreements and the co-ordination of research and development activities. Certainly these policies, coupled with a considerable degree of industrial dynamism, characteristic of Japanese industry in general, have succeeded in placing the industry in a position to successfully challenge a lengthy period of American domination.

A most important recent development in the Far East has been the emergence of a modern semiconductor industry in South Korea, although it is too early at present to assess the effect that this is likely to have on a worldwide scale. What is significant, however, is its present rapid rate of growth, and also its technically advanced nature.

The development of the European semiconductor industry has taken place against a background of increasing American economic penetration within its market, dating from around 1960. A characteristic of the semiconductor industry in Europe is that, in addition to its smaller size, the market is fragmented between the various nation states. A key factor affecting the growth of the industry is the relationship between market size and unit costs and this fact accounts for certain important disadvantages, since a consequence of smaller market size is shorter production runs, leading in turn to lower productivity.[33] American subsidiaries in Europe have considered this the main reason for lower output per employee.[34] Also, owing to the fragmentation of the European market and the smaller size of equipment firms, sales efficiency has been lower in Europe than in the United States, a significant reason being that for an equivalent volume of sales the number of contacts made by an individual salesman must be much greater.[35]

Another important disadvantage which has existed from the beginning of semiconductor production in Europe has been the existence of the technical lag amounting to about two or three years mentioned above. Consequently, it has been extremely difficult to compete directly with American firms, since their unit cost of production for a given device type during this time would have fallen to a much lower level. Also, owing to the rapid rate of innovation within the industry, by the time the European product approaches the point of being cost-competitive, a new American product may well be entering production, rendering the original device obsolete. Reasons for this technical lag have been suggested and include lack of investment, particularly in the area of production, and also lack of suitable markets. Further suggestions have included the low status of production engineering compared with research (at least in the case of Britain), the absence of skill clusters and also inefficient management, particularly in the marketing sector.

Although American firms have frequently made licensing agreements with their European counterparts, there has been a marked reluctance on the part of American companies to provide the European semiconductor industry with research and development expertise, or the means to acquire it. The cost of licencing agreements has imposed an additional financial burden on the industry and is a further consequence of technical lag. No comprehensive data appears to exist regarding the amount of royalties paid by the semiconductor industry in any country, but all firms manufacturing transistors were paying fees to Western Electric for the use of the basic transistor patents as late as 1968 (with the exception of hearing aids, royalties on these being waived by Bell in 1954).[36]

Certainly, lack of direct investment by governments and the relatively small size of European defence procurement contracts have been an important factor in inhibiting the growth of the industry. European semiconductor companies are unlike their American counterparts, being vertically integrated in structure, and have largely financed their semiconductor operations from profits made in other areas of activity. Unlike the situation in the United States, there has been a general lack of available capital from the start. Few new firms devoted entirely to semiconductor manufacture have arisen in Europe and of these, even fewer have been successful on a long term basis. As an example of the relative weakness of the military market in Europe, it has been estimated that, in 1967, the total market value of the Western world's electronic industry was $13·1 billion, and the American share of this figure amounted to over 80%.[37] Unlike the situation in Japan, the European semiconductor industry has not benefited from an alternative source of finance, such as long-term low interest loans from the banking sector of the economy. Financial aid to industry in Europe has been mainly invested in the field of research and development, unlike the situation in both Japan and the United States, where the vital field of device production has not been neglected. Also, financial assistance, at least in the United Kingdom and France, has been made available by government to both indigenous companies and foreign subsidiaries alike.

An important factor affecting European performance in semiconductor production has been the general lack of co-operation and co-ordination between individual manufacturing organisations within the area. Separated by cultural and language barriers and subject to the policies of individual governments, indigenous firms have concentrated mainly on supplying the home market. However, in order to obtain a foothold in the American market and overcome technical lag, a number of semiconductor manufacturing firms within the United States have been purchased by large European organisations; for example Siliconex by Philips and Fairchild by Schlumberger. This approach has also resulted in increasing the availability of American technology. The net overall result has however been somewhat disappointing, the European share of the worldwide semiconductor market having fallen from 16% in 1974 to 11% in 1986 (see Fig. 7.3). It is notable, however, that Philips, with its wide range of overseas investments, including its stake in the growing Japanese semiconductor industry, has recently shown some improvement in performance (see Tables 7.5 and 7.6).

Because of the importance of the Eastern bloc, some mention must also be made of developments in that area. In view of the lack of data regarding the USSR, it is difficult to speculate upon the development of the semiconductor industry in that country, or indeed within the rest of Comecon. Certainly, however, the US embargo on strategic goods, supported by other nations, cannot fail to have had an adverse effect on the initial development of semiconductor technology. Nevertheless, in view of its increasing strategic importance, it is unlikely that this industry will be neglected by the Soviet authorities and it is possible that the present inability of the USSR to obtain both semiconductor devices and electronic equipment on a substantial scale from elsewhere may stimulate greater efforts than might have been made otherwise to become self-sufficient in this field.

It appears from the foregoing analysis that the development of the semiconductor industry within the three major geographical areas concerned has differed significantly. In spite of its increasingly international character, the dominant centre has remained (at least until the middle of the present decade) overwhelmingly within the United States, reflecting the wider aspects of that country's political and economic power and influence.

The recent success of the Japanese semiconductor industry has demonstrated that a viable alternative to the American industrial model not only exists, but may well prove superior in terms of capturing a major share of the international semiconductor market. The major factor which Japanese and American industry appear to have in common, and which Europe does not, is that they have both received considerable capital investment and this investment has been made in the vital production–development sector, in addition to research and development. In the early stages of semiconductor manufacture, the Japanese, unlike the Europeans, operated what amounted in practice to a policy of trade protection, preventing the intrusion of both foreign capital and the excessive import of semiconductor devices. To operate a policy of this nature successfully requires the existence of a substantial and preferably growing home market, and it is doubtful if any one European nation, acting alone, now possesses the resources to adopt this approach. Any possibility of success for the European semiconductor industry appears to lie in the direction of a considerably greater degree of co-operation between individual firms, together with a much larger increase in government support than in the past.

The history of the Japanese semiconductor industry has demonstrated that a fundamental prerequisite for the growth and maintenance of a viable semiconductor industry is the availability of long-term loans at low interest rates. This is of particular importance in an area where financial losses over long periods must be sustained. To ensure a similarly strong European industry, some mechanism needs to be established to ensure such funding on an adequate scale.

The establishment of a common European semiconductor policy would necessarily need to be of a long-term nature, probably extending over at least two decades and therefore critically dependent upon the continuity of a sympathetic and consistent political policy over that period. If inter-European co-operation of this nature cannot be realised, the most likely scenario appears to be a continuing decline in relative position within the world market of the European semiconductor industry, together with increasing foreign domination.

The consequences of the alternative policy of Europe abandoning semiconductor manufacture should be considered. Although this might well release resources which could be made use of in other ways, the price for this decision would be dependence upon those nations manufacturing the key building blocks required by the electronics industry, and effective control of that industry would therefore be lost. Such an event could hardly fail to lead to a situation of rapid technical and economic decline, affecting as it would both the commercial and industrial sectors of the economy. Indeed, looking into the future, it is difficult to envisage a situation in which any nation (or group of nations) having abandoned the struggle to keep abreast of advances in this new world-changing technology, would be able to sustain any real measure of control over its destiny.

Having committed ourselves to a future in which the consequences of the application of semiconductor technology throughout society will become increasingly important, a corresponding awareness of the nature and scale of the changes involved becomes essential. Certainly, developments in society on an unprecedented scale are likely to occur within the next few decades as a direct result of improvements in solid state devices, these developments bringing with them problems of an unforseen nature. Faced with this situation, a need exists for a wide-ranging debate to consider these problems as they arise, involving the skills of the technologist, economist, social scientist and historian. This need is now urgent and efforts should therefore be made to establish links between these disciplines with this common aim in view.

9.3 References

1 *Amer. J. Sci.*, 1983, **26**, p. 465
2 BOSE, J.C.: US Pat. No. 755540, 1904
3 *Phys. Rev.*, 1907, **25**, p. 31 *Ibid.*, 1909, **28**, p. 153 *Ibid.*, 1909, **29**, p. 478
4 GIBBONS, M. and JOHNSON, C.: 'Science technology and the development of the transistor', 'Science in context', (OUP, 1982 p. 183
5 FITTS, C.E.: 'A new form of selenium cell', *Amer. J. Science*, 1883, **26**, pp. 465–472
6 BERGMANN L.: 'On a new selenium barrier photocell', *Phys. 2*, 1931, **32**, pp. 286–288
7 GRONDAHL, L. and GEIGER P.: 'A new electronic rectifier' *Trans. EE*, Feb. 1927, pp. 357–366
8 GIBBONS, M. and JOHNSON, C.: *Ibid.* pp. 179–180
9 MOTT, W.F.: 'The theory of crystal rectifiers', *Proc. Royal Soc.*, 1939, **171**, pp. 27–38
10 SCHOTTKY, W.: 'Zur Halbeiter theorie der Sperrschichtrichter und Spitzengleichrichter', *Zeitschrift fur Physik*, 1939, **113**, 1939, pp. 367–414
11 DAVYDOV, B.: 'On the contact resistance of semiconductors', *J. Phys. (USSR)*, 1939, **1**, 2, 1939, pp. 167–174. Also 'The rectifying action of semiconductors', *Tech. Phys. USSR*, 2, 1938, **5**, pp. 79–86
12 SMITH, R.A.: 'Semiconductors', CUP, 2nd Ed. p. 6
13 SOMMERFELD, A.: *Z. Phys.*, 1928, **47**, p. 1
14 SMITH, R.A.: *Ibid.*, p. 10
15 WILSON, A.H.: 'The theory of electric semiconductors II', *Proc. Royal Soc. A*, 1931, **133**, p. 458 *Ibid.* 1931, **134**, p. 277
16 WILSON, A.H.: *Ibid.*, **134**, p. 277
17 PFANN, W.: 'Zone refining', *Scientific American*, Dec. 1967, p. 63
18 PFANN, W., *Ibid.*
19 WEINER, C.: 'How the transistor emerged', *IEEE Spectrum*, Jan. 1973, p. 26
20 SPARKES, M.: 'The junction transistor', *Scientific American*, July 1952, p. 29
21 SHIVE, J.N.: 'The transistor, a new semiconductor amplifier', *Proc. IEEE*, 1954, **72**, p. 1696
22 BRAUN, E. and MACDONALD, S.: 'Revolution in miniature', (CUP 2nd Ed., 1982), p. 45
23 WEINER, C.: *op.cit.*, p. 32
24 BRAUN, E. and MACDONALD, S.: *op.cit.*, pp. 38–39
25 SMITH, R.A.: *op.cit.*, pp. 8–39
26 BRAUN, E. and MACDONALD, S.: *op.cit.*, p. 88
27 US Dept. of Commerce: 'Semiconductors: US Production and Trade' (US Govt. Printing Office, 1961), pp. 15–16
28 Ref. Fig. 4.3
29 *Ibid.*
30 US Dept. of Commerce: *Ibid*
31 GOLDING, A.M.: D. Phil thesis, University of Sussex, 1971,, p. 137
32 OECD: Gaps in technology series 'Electronic Components' Paris, 1968, p. 57
33 *Ibid.*, p. 77
34 *Ibid.*, p. 110
35 *Ibid.*, p. 39
36 *Ibid.*, p. 111
37 *Ibid.*, p. 26

Glossary of commonly used terms in semiconductor technology

This review is not intended to be comprehensive, but to explain terms commonly used when discussing the technology of semiconductors.

Acceptor

This is an impurity,which, when added to intrinsic semiconductor material, or, for example, to N-type extrinsic semiconductor material in a sufficient amount, makes that material P-type.

Alloyed junction

A junction which has been formed by recrystallisation from the molten state in which there is an abrupt change from P- to N-type material.

Avalanche breakdown

This term describes the phenomenon of reverse biased junction breakdown due to carrier multiplication. It is caused by the application of a reverse voltage of sufficiently high potential to enable a large reverse current to flow. The magnitude of the breakdown voltage is a function of the relative resistivities on either side of the P–N junction.

Bandwidth

This is defined as the range of frequency response of an amplifier over which the power output remains above half its maximum, or mid-frequency value.

Base

This central region of a bipolar transistor is so called because it comprised the wafer upon which the point contacts of the original transistor were located. It is situated betwen the emitter and collector regions of the device.

Bipolar transistor

This device consists of a three layer P–N–P or N–P–N sandwich of semiconductor material, its action being initiated by the injection of majority carriers from the emitter into the base. Here they exist as minority carriers and diffuse towards the collector, the majority then being swept into it by the aiding field provided by the reverse biased base–collector junction. Applications include its use as an amplifier, oscillator, and as an electric switch.

Charge carrier

This is a carrier of electrical charge within a semiconductor crystal. It can exist either as an electron (negative charge) or as a hole (positive charge).

Chip

A portion of a semiconductor slice containing either a single device or integrated circuit.

Collector

This is the region of a bipolar transistor which receives the base transit charge.

Cut-off frequency

The frequency at which transistor gain falls to a certain defined value. For example, the frequency at which the gain of a bipolar transistor in common-emitter configuration is equal to unity. Cut-off frequency is a function of transistor base width, increasing with decrease in base thickness.

Depletion layer (or region)

In this region of a semiconductor junction the charge carriers have been removed by the electric field existing at the junction. The region has therefore been depleted of free charge carriers. The depletion layer acts as a barrier to majority carrier movement, although thermally generated minority carriers are able to cross the barrier because of their opposite charge polarity.

Die (plural dice)

See chip.

Dopant

An element which, when added to a semiconductor crystal, strongly increases its ability to act as an electrical conductor.

Dynamic memory

This type of memory will retain information without the need of a periodic re-write cycle. The data will remain in store as long as power is applied.

Emitter

This region of a bipolar transistor acts as a source of mobile charge carriers which it injects into the base region.

Hole

This term describes the absence of an electron in the covalent band of an atom within a semiconductor crystal. The absent electron acts as a positive charge, consequently hole conduction in semiconductors is equivalent to the motion of positive charges.

Integrated circuit

An array of circuit elements, typically diodes, transistors and possible other circuit parts, formed during the same manufacturing process on a single substrate and interconnected to form an electronic circuit. The silicon planar method of device construction lends itself readily to this type of approach.

Intrinsic conduction

This is the process of electrical conduction which would take place in a pure, or intrinsic, semiconductor material; namely, one that has no impurities present.

Ion implantation

This method of forming P–N junctions has now to some extent supplanted the diffusion process, having the advantage of more accurate definition of the dimensions of P–N junctions. This factor is of great importance in the fabrication of integrated circuitry. High-voltage ionic bombardment is used to introduce the selected impurities into the desired regions. A wider range of impurity gradient can be obtained by the use of this process than can be achieved using the alternative method of diffusion.

Junction

This is the boundary between two semiconductor regions usually of P-type and N-type conductivity. A P-type and an N-type semiconductor in electrical contact constitutes a junction diode.

Logic gate

An electrical circuit in which an output having constant amplitude is registered if a particular combination of input signals exists. Examples are 'OR, AND, NOT & INHIBIT' circuits. These logic functions may be realised by various bipolar families, for example, diode transistor logic (DTT), transistor–transistor logic (TTL) and emitter coupled logic (ECL) or by MOS & CMOS logic.

Majority carrier

This is the mobile charge carrier most responsible for electrical conduction (either a hole or an electron) within a semiconductor material. Thus, electrons in N-type are majority carriers in N-type material and holes in P-type material.

Minority carrier

This is the mobile charge carrier of opposite electrical polarity to the majority carrier (either a hole or an electron) which exist within a semiconductor material. Thus, holes are minority carriers in N-type material and electrons in P-type material.

N-type material

Semiconductor material in which the majority carriers are electrons.

Ohmic contact

A semiconductor-to-metal contact, or a metal-to-metal contact which does not behave as a rectifier.

Oxide masking

The process of using an oxide to protect a chosen area of a semiconductor surface during diffusion. Using this technique, selected impurities may be diffused through 'windows' etched through the oxide in order to form P–N junctions.

Passivation

The protection of the surface of a semiconductor in order to ensure that its electrical properties remain stable. This is achieved in the planar process by growing a layer of silicon dioxide over the junctions of the device, electrically stabilising its surface and shielding it against moisture and contamination.

Photoresist

This is a photosensitive material, which, after exposure to ultra-violet light, resists the action of an etch, for example, hydrofluoric acid. Photoresist is widely used in the planar process to carry out the process of selective etching of silicon surfaces.

Polycrystalline

This term is used to describe a material consisting of many small distinct crystalline regions of different orientation, rather than a single crystal. Polycrystalline semiconductor material is unsuitable for fabricating transistors, owing to the grain boundaries between each crystal acting as recombination centres for charge carriers.

P-type material

Semiconductor material in which the majority carriers are holes.

Random access memory (RAM)

This is a memory device in which the access time is the same for any address within the memory.

Recombination

The mechanism whereby holes and electrons recombine. It is affected both by surface imperfections and volume impurities within the crystal. Recombination in the base of a bipolar transistor reduces the number of minority carriers available for conduction, thus reducing current gain.

Rectifying contact

An electrical contact, permitting the easy flow of current in one direction (called the 'forward' direction) whilst restraining the flow of current in the opposite direction (the 'reverse' direction).

Semiconductor

This is an element intermediate in electrical conductivity between insulators (nonconductors) and metals (conductors). The most widely used semiconductor in electronic devices at present is silicon.

Slice

A thin slab of semiconductor sawn from an ingot. The ingot is usually in single crystal (monocrystalline) form, and may be currently up to six or eight inches across, although, in the early days of semiconductor manufacture, they were typically half an inch in diameter. The advantage of larger diameter slices is economy of scale in production. In the planar process, discrete devices or integrated circuits are manufactured on the silicon slice, often several thousand or more per slice. The slice is cut up into individual dice before final assembly and encapsulation.

Static memory

This type of memory will retain information without the need of a periodic re-write cycle. The data will remain in store as long as power is applied.

Substrate

The underlying layer of material upon which a device, circuit or epitaxial layer is fabricated. In present semiconductor manufacture this layer is usually a silicon slice.

Surface states

The electrical properties of a semiconductor are strongly influenced by its surface conditions. This is because of the interaction between the environment and the particular electrical conditions present at the semiconductor surface. A problem in device manufacture is that these surface states may cause electrical instability and high surface leakage currents. The planar process overcomes this problem by electrically stabilising or passivating the surface by means of a layer of silicon dioxide.

Unipolar transistor

This is a device such as a field effect transistor whose action depends upon the movement of majority charge carriers only.

Very large scale integration (VLSI)

This is defined as a chip containing one hundred thousand or more components.

Wafer

This is a slice of semiconducting material containing perhaps many dice (or chips). Each chip may contain, for example, a single device or an integrated circuit. The wafer is separated into individual dice during a latter stage of the production process.

Zener breakdown

This type of reverse breakdown voltage takes place when the junction depletion layer is very narrow; consequently a high electric field intensity exists across the junction and may be sufficient to tear electrons directly out of the covalent bonds. The hole–electron pairs so formed contribute to the reverse leakage current, causing voltage breakdown at values of below about 6 V in silicon. The breakdown effect is reversible and not damaging if power loss in the diode is limited by some means, and the effect may be used to design voltage stabilisers with given values of breakdown voltage.

Basic semiconductor junction theory

The following qualitative account of rectification and transistor operation is not intended to be comprehensive, but has the dual purpose of giving a very basic outline of these processes and defining the terms used in the preceding chapters.

A semiconductor material is one which is defined as having a resistance somewhere between that of resistors and insulators. Examples of these are silicon and germanium (Group IV elements in the periodic table). In addition to the basic elements, complicated semiconductor structures can be manufactured by mixing individual elements. Semiconductors can be classified according to the particles they carry. There are two types: firstly, ionic, where the current is carried by ions (this action results in change of chemical composition), and secondly electronic, where conduction is by electron flow. In this case, no change in chemical composition occurs. Only the latter type are used in semiconductor device fabrication.

Electronic semiconductors are classified in turn as either *P or N type*. In N type semiconductors the predominant charge carriers are *electrons* (negative charge carriers) and in P type semiconductors a deficit of electrons exists, and the material behaves as if charges equal and opposite in sign to the electrons exist. These charges are called *holes*. When an electric field is applied, holes and electrons move in opposite directions, owing to the difference in polarity of charge.

Some semiconductors are P type only, for example copper oxide, whilst others are N type only, for example zinc oxide. Silicon and germanium can be made either P or N type by adding different impurities in very small amounts. If germanium or silicon could be purified sufficiently it would be neither P nor N type. In this impurity-free state the semiconductor material is called *intrinsic*. *Extrinsic* material is that to which either P- or N-type impurities have been added. The addition of the impurities is termed *doping*. By adding minute quantities of impurities, the characteristics of the semiconductor materials can be greatly modified.

A rectifying semiconductor may be made in two ways, either by making a suitable contact between a semiconductor and a metal (this type of rectifier being known as the 'cat's whisker' or *point-contact diode*), or between two semiconductors, forming a 'P–N junction', this type usually being referred to as a *junction diode*. These rectifying devices may be used for modulating, detecting and rectifying at low current levels. The junctions so formed may exhibit photoelectric properties, for example *photoconductivity*. In this case the electrical resistance of the semiconductor element decreases when light falls upon it, allowing an increase in current to flow in an external circuit, or the *photovoltaic effect*. In this case, a voltage is generated when light falls on a rectifying contact or junction.

When a P–N junction is formed, holes in the P material and electrons in the N material attempt to cross it. This results in oppositely charged immobile ions being 'uncovered' on opposite sides of the junction. A consequence of this is that an electric field is set up at the junction, which opposes any further movement of charge carriers across it. Since this barrier set up is depleted of free charge carriers it is called a *depletion layer* or alternatively a *space-charge layer*. These free charge carriers (i.e. holes in the P type material and electrons in the N-type material) are called *majority carriers*. In most semiconductor junctions, the barrier is a fraction of a millimetre thick (typically 10^{-4} cm). At the same time, owing to thermally generated atomic lattice vibration within the semiconductor material, 'electron–hole pairs' are generated on both sides of the junction. The holes thus generated in the N-type material and the electrons in the P-type material are called *minority carriers*. These temperature-dependent minority carriers, owing to their charge polarity, are swept across the junction by the field at the depletion layer, which acts in such a way as to aid this action, owing to its polarity. At a given temperature and with no externally applied voltage, equilibrium is established between the movement of charge carriers at the depletion layer.

Now let a P–N junction be connected to an external battery in such a way that the P material is biased positively with respect to the N material. The applied potential exerts a force opposing the field of the depletion layer, causing it to become narrower, and, for a given temperature, allowing more majority carriers to cross the junction. Under these conditions, only the thermally generated minority current flows across the junction. (In practice a further current contribution will occur owing to effects at the semiconductor surface.) This current is commonly referred to as *reverse leakage current*. The junction is now operating in the reverse blocking mode, and this condition is called *reverse bias*. If the reverse voltage is sufficiently great the gradient of the electric field of the junction will become high enough to cause the formation of the hole–electron pairs in the vicinity of the junction and the reverse leakage current will rapidly increase for little further increase in applied voltage. This condition is called *reverse breakdown* and the voltage at which this effect occurs is termed *reverse breakdown voltage*.

Historically, the *point-contact transistor* was invented before its junction counterpart, although it was rapidly supplanted by the latter device. It consists of two closely spaced metal probes in contact with a slab of semiconductor material, thereby forming two adjacent metal-to-semiconductor contacts. The theory of the junction transistor is less complex and easier to explain. For these reasons what follows is a basic account of the functioning of the junction transistor.

The *junction transistor* may be considered as either an N–P–N or P–N–P sandwich of semiconductor material, one junction being forward biased with respect to the central section, termed the base, and the other junction reverse biased. Consider the N–P–N type. The N region of the forward biased junction is called the emitter, and the N region of the reverse biased junction is called the collector.

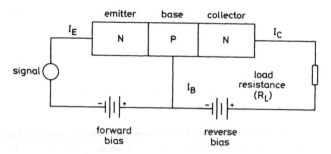

Fig. B.1 Action of the N–P–N junction transistor

The action of the device is as follows. The forward biased emitter junction injects majority carriers into the P-type base, where they become minority carriers. These minority carriers diffuse across the base and then, owing to their charge polarity, are swept by the field at the reverse-biased collector into the collector region. If the base is narrow, most of the emitter current (in the form of electrons) will reach the collector. However, if the base width is increased in thickness, this ratio will decrease, owing to hole–electron pair combination in the base. Successful transistor action depends upon this ratio being high, since, at some definite thickness in base width, transistor action will completely cease. Under normal operating conditions the collector current is determined mainly by the emitter current rather than the collector voltage, and because of this, the transistor described is referred to as a *current controlled device*. Also, a very small emitter voltage is capable of producing a large emitter current owing to the low resistance of the forward-biased emitter–base junction. Therefore, if a small signal is superimposed upon the emitter current, flowing in the low-resistance emitter circuit, it controls a current of almost equal size flowing in the reverse-biased high-resistance collector–base circuit. An external load resistance R_L may now be connected in the collector circuit and its value made high compared with the low resistance in the emitter circuit; consequently more signal power will be developed across R_L than that developed in the emitter circuit, thus achieving power amplification. The increase in signal power is of course obtained at the expense of the direct current supply.

The electrical connection described is called the 'common base' configuration since the base terminal is common to both input and output. It is also possible to obtain transistor action in the common-emitter and common-collector configurations. Depending upon the type of circuit configuration, voltage, current and power gains can be obtained.

In order to successfully deal with high frequencies, the base of the transistor must be extremely thin to reduce the transit time of charge carriers across it to a minimum. In addition, the base resistance should be as low as possible and the collector depletion layer capacitance should be minimised. High power devices, however, require wider bases and therefore perform less well at high frequencies. The design of transistors is inevitably a compromise between conflicting factors.

Both minority and majority carriers are involved in transistor action of the type described and these devices are referred to as *bipolar* transistors, to distinguish them from unipolar types, which rely on the action of only one type of charge carrier. The *field-effect* transistor is an example of the unipolar type. Field effect transistors, due to the way they are constructed, have a very high input resistance and virtually no current flows into the device. Control is achieved by varying the input voltage, and this device is consequently referred to as a *voltage controlled* device and has this characteristic in common with the thermionic valve.

For further information on this topic, including an outline of the physical principles forming the basis of junction transistor theory, the following books are recommended to the reader:

ADLER R.B., SMITH A.C., and LONGI R.L. 'Introduction to semiconductor physics'. Vol 1, (J. Wiley, 1964
FRASER D.A. 'The physics of semiconductor devices'. (Clarendon Press, 1977
KEMP B. and McDONALD R.H.: 'Fundamentals of modern semiconductors, (Foulsham, 1965)
WRIGHT H.C. 'Elementary semiconductor physics' (Van Nostrand Reinhold, 1978)

The physical modelling upon which the above description of transistor action rests has not remained static, but is in the process of continual development. For an account of recent fundamental work, presented with a minimum of mathematics, the reader is referred to the following publication:

HEY T. and WALTERS P. 'The quantum universe' (CUP, 1987)

Pattern of a typical semiconductor product cycle

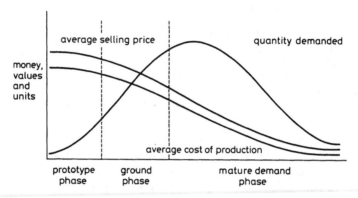

Fig. C The pattern of a typical semiconductor product cycle (Source: GOLDING, R.H.: D. Phil. Thesis, University of Sussex, 1971, p. 92)

Characteristics and applications of gallium arsenide

Although the development of the semiconductor industry has largely been dominated by the elements germanium and silicon, some mention must be made of the compound semiconductor gallium arsenide (GaAs) which has been widely used for high frequency, high speed switching and high temperature applications. Military requirements in particular have given work in the microwave field a strong and continuing stimulus and the considerable efforts expended in the development of high frequency devices fabricated from gallium arsenide should largely be seen within this context.

The characteristics of gallium arsenide are such that it performs well as a solar cell and light emitter, and its energy bond structure makes possible transferred electron (Gunn) oscillations. Because of its high electron mobility and saturation drift velocity, this material offers distinct advantages in the field of high frequency operation. In the intrinsic state, it has a high resistivity at normal temperatures and therefore components made on an intrinsic substrate may not need an isolation diffusion. Also, owing to its greater energy bond gap than silicon, operation at higher temperatures is theoretically possible. Unfortunately, this material is difficult to process and therefore production costs are much higher than in the case of silicon, and this fact has certainly inhibited its much wider use in the past. One problem is that it does not grow an electrically stable oxide layer in the same way as silicon, since one element oxidises more rapidly than the other, leaving a metallic phase at the interface, thus rendering planar processing impossible. Since arsenic evaporates from the melt and the crystal above about 600°C, diffusion techniques cannot be used. Furthermore, gallium arsenide is difficult to dope and crystal defects tend to be higher than is the case for silicon. Consequently, techniques for fabricating devices from this material have taken much longer to develop than was the case for both silicon and germanium.

High frequency devices manufactured from gallium arsenide include the Gunn diode (developed by J.B. Gunn at the Radio Corporation of America, and first fully explained by A. Chyroweth, of Bell Telephone Laboratories in 1965)[1] and the Impact Ionisation Avalanche Transit Time (IMPATT) diode (described by S.M. Sze and R.M. Ryder in 1971).[2] Both IMPATT and Gunn diodes can be operated in the negative conductance mode to provide oscillations and amplification. The IMPATT diode is one of the most powerful sources of power at microwave frequencies whilst the Gunn diode operates at very low noise levels. Gallium arsenide tunnel diodes operating in the gigahertz region have been produced from the early 1960s onwards. Their applications include use as microwave oscillators and high speed switches. Field effect transistors, made from this material, both of the junction and metal–semiconductor type, are in production and have been used as replacements for travelling wave tubes at lower powers. They have also replaced Gunn diodes to some extent up to frequencies of about 10 GHz, although their noise levels are somewhat higher. Ultra-high-speed gallium arsenide integrated circuits are currently being manufactured in Japan, the United States and Europe and most types are now compatible with silicon bipolar emitter coupled logic (ECL). The former material has been used in the manufacture of solar cells and also lasers, which were first developed in 1963.[3] When used in laser applications, the output of GaAs is in the infra-red. (However, for work in the visible spectrum, doping with phosphorus causes a shift in frequency response in the required direction.) Devices have been developed for computer applications up to the microprocessor level, and custom, semi-custom and gate array circuitry has been implemented. A further application has been in the field of fibre-optical transmission systems.

Recent work with gallium arsenide both in the United States and Japan has resulted in the production of devices which are beginning to challenge and outperform silicon in an increasing number of applications. For example, in 1988 Toshiba stated that they were sampling a 10 W gallium arsenide chip for microwave communication to operate in the frequency range 8·5 to 14·5 GHz[4]. Also in that year Triquint Semiconductor (the former gallium arsenide division of Tetronix) claimed that they were manufacturing low dislocation density material, and achieving yields approaching silicon, using 4 in diameter wafers.[5]

Improvements in material quality of this nature are a firm indication that gallium arsenide will play an increasingly larger role in the future semiconductor industry.

References

1 VOELCKER, J.: 'The Gunn effect' *IEEE Spectrum*, July 1989, p. 24
2 SZE, S.M. and RYDER R.M.: *Proc. IEEE*, 1971, **50**, (40)
3 HALL, R.N., *et al.*: *Phys. Rev. Letters*, 1962, **9**, pp. 366–368
4 WICKNAMANAYAKE, D.: *Electronics Weekly*, 2 Nov. 1988
5 MANNERS, D.: *Electronics Weekly*, 6 July 1988

Bibliography

Public Documents

'Hungary: Market Report' (Published by British Embassy, undated)
US Dept. of Commerce: 'A Report on the Semiconductor Industry' (US Government Printing Office, Washington, DC, 1979)
US Dept. of Commerce: 'Semiconductors: US Production and Trade' (US Government Printing Office, Washington, DC, 1961)·
US International Trade Commission: 'Competitive factors influencing world trade in integrated circuits' (US Government Printing Office, Washington, DC, 1979)

Books

ADLER, R.B., SMITH, A.C. and LONGI, R.L.: 'Introduction to Semiconductor Physics. Vol. 1' (J. Wiley & Sons, 1964)
AMANN, R.: 'Technical progress and Soviet economic development' (Blackwell, 1986)
ANTEBI, E.: 'The electronic epoch' (Van Nostrand, 1982)
ATHERTON, W.A.: 'From compass to computer' (San Francisco Press, 1984)
BRAUN, E. and MACDONALD, S.: 'Revolution in miniature' (Cambridge University Press, 2nd ed., 1982)
DONALDSON, P.: 'Guide to the British economy' (Penguin Books, 1965)
DUMMER, G.W.A.: 'Electronic inventions 1745–1956' (Pergamon Press, 1977)
DUMMER, G.W.A.: 'Fixed resistors' (Pitman, 2nd edn., 1967)
FRASER, D.A.: 'The physics of semiconductor Devices' (Clarendon Press, Oxford, 1977)
READ, J.W. (ed.): 'Gate arrays', (Collins, 1985)
GIBBONS, M. and JOHNSON, C.: 'Science technology and the development of the transistor' in 'Science in context'. by Barnes, B. and Edge., E. (Eds.): (OUP, 1982)
GREGORY, G.: 'Japanese electronics technology: Enterprise and innovation' (Japan Times Ltd, Tokyo, Japan, 1985)
HEY, T. and WALTERS, P.: 'The quantum universe' (CUP, 1987)
JEWKES, J., SAWERS, D. and STILLERMAN, R.: 'The sources of Invention' (Macmillan, 2nd edn., 1969)
KEMP, B. and McDONALD, R.H.: 'Fundamentals of modern semiconductors' (Foulsham, 1965)
RUDENBURG: 'World semiconductor industry in transition 1978–83' (Arthur D. Little, Cambridge, Mass., 1980)
SERVAN-SCHREIBER, J.: 'The American challenge' (Penguin Books, 1967)
SMITH, R.A.: 'Semiconductors' (CUP, 2nd edn., 1978)
STARLING, S.E. and WOODALL A.J.: 'Physics' (Longmans, 2nd ed., 1961)
SZE, S.M.: 'VLSI technology' (McGraw–Hill, 1983)
WRIGHT, H.C.: 'Elementary semiconductor physics' (Van Nostrand Reinhold, 1978)

Articles

ABEGGLEN, J.C.: 'The economic growth of Japan' *Scientific American*, March 1970, p. 31
ADAMS, W.E. and DAY, R.E.: 'The action of light on selenium', *Proceedings of the Royal Society of London*, 1876, **25**, pp. 113–117

ADCOCK, W.A., *et al.*: 'Silicon transistor', *Proc. IRE*, 1954, **42**, p. 1192

ALEKSEEV, N.T. and MALAINOV, D.C.: 'Generation of high power oscillations with a magnetron in the centimetre band', *Journal of Tech. Phys. (USSR)*, 1940, **10**, 1940, pp. 1297–1300

ALGER, J. and CLAYTON, Sir R.: 'Electronics in the GEC Hirst Research Centre–The first 60 years', *Radio & Electronic Engineer*, 1984, **54**, p. 305

APPLETON, Sir E.: 'Thermionic devices from the development of the triode up to 1939'. Lecture delivered before the IEE, 16th November 1954. In 'Thermionic valves, 1904–54' (IEE, 1955) pp. 17–25

ARNQUIST, W.N.: *Proceedings of the IRE*, 1959, **47**, pp. 1420–1439

ASCHNER, J.F., *et al.*: 'A double diffused silicon high frequency transistor produced by oxide masking techniques', *J. Electrochemical Soc.*, 1959, **106**, pp. 1145–1147

BAEDEKER, K.: 'Using Cu I', *Phys. Zeitung*, 1909, **29**, p. 506

BARDEEN, J.: 'The improbable years', *Electronics*, 19th February 1968, p. 78

BATES, J.J. and COLYER, R.E.: 'Electrical power engineering', *Radio & Electronic Engineer*, 1973, **43**, pp. 116–117

BEALE, J.R.A.: 'Alloy diffusion: A process for making diffused base junction transistors' *Proc. Phys. Soc. B*, 1957, **70**, p. 1087

BELLO, F.: 'The year of the transistor', *Fortune*, March 1953

BEQUEREL A.E.: 'On electric effects under the influence of solar radiation', *Comptes Rendues de l'Academia des Sciences*, 1839, **9**, pp. 711–714

BERGMANN, L.: 'On a new selenium barrier photocall', *Phys Z.*, 1931, **32**, pp. 286–288

BOOT, A.A. and RANDALL, J.T.: 'The cavity magnetron', *Journal IEE*, 1946, **93**, Part IIIA, pp. 928–938

BOSTOCK, G. (Mullard Ltd.): 'Bipolar and high speed arrays', in READ, J.W. (Ed.): 'gate arrays' (Collins, 1985), p. 64.

BOTTOM, V.E.: 'Invention of the solid state amplifier', *Physics Today*, May 1964, pp. 60–62

BOWER, R.W., *et al.*: 'M15 field effect transistors formed by gate masked ion implantations', *IEEE Trans.*, 1968, **ED–15**, pp. 757–761

BRAUN, F.: 'Resistance polarity in metal sulphides' *Ann Pogg.*, 1874, **153**, p. 556

BRIDGES, J.M.: 'Integrated electronics in defence systems', *Proc. IEEE*, Dec. 1964, p. 1407

BROTHERS, J.S.: 'Integrated circuit development', *Radio & Electronic Engineer*, 1973, **43**, p. 43

BROWN, W.L.: *Phys. Review*, 1953, **91**, p. 518

CASE, T.W.: 'Thalofide cell—A new photo–electric substance', *Phys. Review*, 1920, **15**, pp. 289–292

COHEN, J.M.: 'An old timer comes of age', *Electronics*, 19th Feb. 1968

DAVYDOV, B.: 'On the photo electromotive force in semiconductors', *Technical Physics of the USSR*, 1938, **5**, pp. 79–86

DAVYDOV, B.: 'The rectifying action of semiconductors', *Technical Physics of the USSR*, 1938, **5**, pp. 87–95

DE FOREST, L.: 'The Audion: a new receiver for wireless telegraphy', *Trans. American IEE*, 1906, **25**, p. 735

DETTMER, R.: 'Prophet of the integrated circuit', *Electronics & Power*, April 1984, p.287

DOUGLAS, R.W. and JAMES, E.G.: 'Crystal diodes', *Proceedings IEE*, 1951, **98**, Part III, p. 160

DUMMER, G.W.A.: 'Integrated electronics development in the UK and W. Europe', *Proc. IEEE*, December 1964, p. 1415

EARLY, J.M.: 'Semiconductor devices', *Proceedings IRE*, 1962, **50**, pp. 1006–1010

EINSTEIN, A.: 'A heuristic standpoint concerning the production and transformation of light', *Ann. Physik*, 1905, **17**, pp. 132–148

FARADAY, M.: 'Experimental researches in electricity' (Bernard Quartier, London.) 1833, **1**, pp. 122–124

FITTS, C.E.: 'A new form of selenium cell', *American Journal of Science*, 1883, **26**, pp. 465–472

FORD, A.W.: 'Brushless generator for aircraft—a review of current developments', *Proc. IEE*, 1962, **109**, pp. 437–55

FROSCH, C.J. and DERRICK, L.: 'Surface protection and selective masking during diffusion in silicon', *J. Electrochemical Soc.*, 1957, **104**, p. 547

FULLER, C.S. and DITZENBURGER, J.A.: 'Diffusion of donor and acceptance elements in silicon', *J. Applied Phys.*, 1956, **27**, p. 544

GEE, C.C.: 'World trends in semiconductor developments and production', *British Communications and Electronics*, 6th June 1959, pp. 450–451

GOLDING, A.M.: 'The semiconductor industry in Britain and the United States: a case study in innovation, growth and diffusion of technology', *D.Phil Thesis*, University of Sussex, 1971

GREENBAUM, W.H.: 'Miniature radio amplifier', *Journal of Audio Engineering Society*, 1967, **15**, p. 442

GRINICH, V.H. and HOERNI, J.A.: 'The planar transistor family', Colloque International sur les dispositifs semiconductors, Paris, February 1961

GRUNDAHL, L.O. and GEIGER, P.H.: 'A new electronic rectifier', *Trans. American IEE*, 1927, **46**, pp. 357–366

GUDDEN, B.: 'On electrical conduction in semiconductors', *Sitzungsberickle der Physmediz Suz. Erlangen*, 1930, **62**, pp. 289–302

HALL, E.H.: 'On a new action of the magnet on electrical currents', *American Journal of Mathematics*, 1879, **2**, pp. 287–291

HEATON, R.: 'Some user notes on sem–custom logic', *in* READ, J.W. (Ed): 'Gate arrays' (Collins, 1985) p. 313

HIBBERD, R.G.: *Proceedings IEE*, Part III, 27th May 1959, pp. 264–278

HIBBERD, R.G.: 'Transistors and associated semiconductor devices', *IEE*, Paper 2914, May 1959

HILSCH, H.R. and POHL, R.W.: 'Control of electron currents with a 3-electrode crystal and as a model of a blocking layer', *Zeits. f. Physics*, 1938, **III**, pp. 399–408

ISTED G.A.: 'Gucliamo Marconi and communication beyond the horizon; a short historical note', *Proceedings IEE*, Jan. 1958, pp. 79–83

JOHNSON, J.B.: Article (unnamed), *Physics Today*, May 1964, pp. 60–62

JONES, R.V.: 'Infra–red detectors in British Air Defence' *in* 'Infra–red physics: Vol. 1' (Pergamon Press, 1961) pp. 153–162

KELLY, M.: *Proceedings of the Royal Society of London*, Series A, 1950, **203a**, pp. 187–301

KEHOE, L.: 'US chipmakers face extinction', *Financial Times*, 6th December 1985

KHANG, D. and ATALLA, M.M.: 'Silicon–silicon dioxide field induced surface devices', IRE/AIEE Solid-State Res. Conference, Carnegie Institute of Technology, Pittsburgh PA, 1960

KILBY, J.S.: *IEEE Trans*, 1976, **ED23**, pp. 648–654

KOMPFNER, R.: 'The travelling–wave tube as an amplifier of microwaves', *Proceedings IRE*, 1947, **35**, p. 124

LAW, R.R., *et al.*: 'A developmental germanium PNP junction transistor', *Proceedings IRE*, 1962, **40**, pp. 1358–1360

LEE, C.A.: 'A very high frequency diffused base germanium transistor', *Bell System Technical Journal*, 1956, **35**, p. 23–24

LEITH, F.A., KING, W.J. and McNALLY, P.: 'High energy implantation of materials'. Final Report, Ion Physics Corp., AD651313, Jan. 1967

LIGENZA, J.L. and SPETZNER, W.G.: *J. Phys. Chem. Solids*, 1960, **14**, p. 131

LOEBNER, E.E.: 'Subhistories of the light emitting diode', *IEEE Trans.*, 1976, **ED23**, pp. 675–698

LOON, H.H., *et al.*: 'New advances in diffused devices'. Presented at the IRE/AIEC Solid–State Device Research Conf., Pittsburgh, PA , June 1967

MACKINTOSH, I.M.: 'The future structure of the semiconductor industry', *Radio & Electronic Engineer*, 1973, **43**, 1973, p. 150

MCLEAN, M.: 'The Koreans are coming', *Electronics Times*, 11th December 1986, p. 20

MOLL, J.L.: *Proceedings IRE*, 1958, **46**, p. 1076

MOLL, J.L. *et al.*: 'P–N–P–N transistor switches', *Proc. IRE*, 1956, **44**, pp. 1174–1182

MORTON, H.L.: 'The present status of transistor development', *Bell System Technical Journal*, May 1952, p. 435

OBERCHAIN, I.R. and GALLOWAY, W.J.: 'Transistors and the military', *Proceedings IRE*, 1952, **40**, pp. 1287–1288

O'CONNOR J.: 'What can transistors do?', *Chemical Engineering*, 1952, **59**, 1952, p. 156

OLSON, K.H. and ROBILLIARD, T.R.: 'New miniature glass diodes', *Bell Laboratories Record*, 1967, **45**, pp. 13–15

OVERTON, B.R.: 'Transistors in television receivers', *Journal of the Television Society*, 1958, (8), p. 444

PAGE, R.M.: 'The early history of radar', *Proceedings IRE*, 1962, **50**, pp. 1232–1236

PARRY, S.: 'Time to get tough with Japan', *Electronics Times*, 19th February 1987, p. 1

PEARSON, G.L. and BRATTAIN, W.H.: 'History of semiconductor research', *Proceedings IRE*, 1955, **43**, pp. 1794–1806

PEARSON, G.L. and SAWYER, B.: 'Silicon P–N junction alloy diodes', *Proceedings IRE*, 1952, **40**, pp. 1348–1351

PENN, T.C.: 'Forcast of VLSI processing—A historical review of the first dry-processed IC', *IEEE Trans.*, 1979, **ED–26**, p. 640

PETRITZ, R.L.: 'Contributions of materials technology to semiconductor devices', *Proceedings IRE*, 1962, **50**, p. 1025

PFANN, W.G.: 'Zone refining', *Scientific American*, December 1967, p. 63

PFANN, W.G.: 'Principles of zone refining', *Transactions of the American Institute of Mining and Metallurgical Engineering*, July 1952, p. 190

PIERCE, G.W.: 'Crystal rectifiers for electric currents and electrical oscillations', *Physical Review*, 1907, **25**, pp. 31–60

PIERCE, J.R.: 'History of microwave tube art', *Proceedings IRE*, 1962, **50**, p. 979

'Reducing internal dimensions improves silicon transistor performance', *Bell Labs. Record*, 1967, **45**, pp. 13–15

ROBERTS, D.H.: 'Silicon integrated circuit', *Electronics & Power*, April 1984, p. 282

ROKA, E.G., BUCK, R.E. and REILAND, G.W.: 'Developmental germanium power transistors', *Proceedings IRE*, 1954, **42**, pp. 1247–1250

ROSENCHOLD, M.A.: 'Experiments on the electrical conduction of solids', *Ann Pogg.*, 1835, **34**, p. 437

ROUALT, C.L. and HALL, G.N.: 'A high voltage, medium power rectifier', *Proceedings IRE*, 1952, **40**, pp. 1519–1521

SABY, J.S.: 'Fused impurity P–N junction transistors', *Proceedings IRE*, 1962, **40**, pp. 1358–1360

SANGSTER, R.: 'Announcing the transistor', 25th Anniversary Observance Transistor Radio and Silicon Transistor, TI Inc., pp. 1–2

SCAFF, J.H. and OHL, R.S.: 'Development of silicon crystal rectifiers for microwave radar receivers', *Bell Syst. Tech. J.*, 1947, **26**, p. 1

SEDUCCA, B.: 'String of poor results for big companies', *Financial Times Survey*, 6th December 1985

SEDUCCA, B.: 'Agressive drive into mass markets', *Financial Times Report*, 6th December 1985

SEIDEL, T.E.: 'Ion implantation' in SZF, S.M. (Ed.): 'VLSI Technology' (McGraw-Hill) p. 260

SHEPHERD, A.A.: 'Semiconductor developments in the 1960s', *Radio & Electronic Engineer*, 1973, **43**, (1/2)

SHEPHERD, A.A.: 'The properties of semiconductor devices', *J. Brit IRE*, 1957, **17**, pp. 262–273

SHOCKLEY, W.: '25 Years of transistors', *Bell Labs. Record*, December 1972

SHOCKLEY, W.: 'The history of P–N junctions in semiconductors and P–N junction transistors', *Bell System Technical Journal*, 1948, **28**, p. 435

SHOCKLEY, W.: 'The path to the conception of the junction transistor', *IEEE Trans.*, 1976, **ED–23**, p. 599

SMITH, W.: 'The action of light on selenium', *Journal of the Society of Telegraph Engineers*, 1873, **2**, pp. 31–33

SNELL, P.: 'Soviet microprocessors and microcomputers', *Technical Progress and Soviet Economic Development*, (Blackwell, 1986) p. 5

'Solid State', *IEEE Spectrum*, Jan. 1987, p. 45

SPARKES, J.J.: 'The first decade of transistor development', *Radio & Electronic Engineer*, 1973, **43**, pp. 8–9

SPARKES, M.: 'The junction transistor', *Scientific American*, July 1952, p. 32

STEPHEN, J.: 'Ion implantation in semiconductor device technology', *Radio & Electronic Engineer*, 1972, **42**, (6)

STEUTZER, O.M.: *Proceedings IRE*, 1952, **40**, pp. 1529–1536

SWART, P. and PARRY S.: 'Japan to pay the price for dumping chips in the US', *Electronics Times*, 20th March 1986

TANENBAUM, M. and THOMAS, D.E.: 'Diffused emitter and base silicon transistors', *Bell System Technical Journal*, 1956, **35**, pp. 1–22

TEAL, G.K.: 'Announcing the transistor'. 25th Anniversary Observance Transistor Radio and Silicon Transistor, TI Inc., 17th March 1980

TEAL, G.K.: 'Single crystals of Ge and Si basic to tbe transistor and integrated circuit', *IEEE Trans.*, 1976, **ED–23**, p. 624

TEAL, G.K.: 'Some recent developments in silicon and germanium materials and devices', Presented at the National Conference on Airborne Electronics, Dayton, Ohio, 10th May 1954
THORNTON, C.G. and ANGELL, J.B.: 'Technology of micro–alloy diffused transistor', *Proceedings IRE*, 1958, **46**, pp. 1166–1176
TOBIAS, J.R.: 'LSI/VLSI building blocks', *IEEE Computer*, 1981, **14**, pp. 83–101
UDA, S., SEKI, T., HATAKEYAMA, K. and SATO, J.: 'Duplex radio telephony with ultra–short waves', *Journal of the Institute of Electronic Engineers of Japan*, 1931, **51**, p. 449
VOGELMAN, J.H.: 'Microwave communication', *Proceedings IRE*, 1962, **50**, p. 907
WALLICH, P.: 'US semiconductor industry: Getting it together', *IEEE Spectrum*, April 1986, pp. 77–78
WALLMARK, J.T.: 'The field effect transistor—An old device with new promise', *IEEE Spectrum*, March 1964, p. 183
WANLASS, F.M. and SAH, C.T.: 'Nanowatt logic using field effect metal oxide semiconductor triodes', *ISSCC Digest*, February 1963, pp. 32–33
WEINER, C.: 'How the transistor emerged', *IEEE Spectrum*, Jan. 1973, p. 28
WILSON, A.H.: 'Theory of electric semiconductors', *Proceedings of the Royal Society*, 1931, **133**, p. 458; and **134**, p. 277
WOHL, R.: 'Soviet research and development', *Defense Science and Electronics*, Sept. 1983, p.11

Reports

'Annual Update of Microprocessors'. EDN, 5th Nov 1980
'Elorg' No.1 (11)' (V/O Vneshtorgreklama, Moscow, USSR, 1986)
'Electronics in Japan' (Electronics Association of Japan)
Ferranti News (Ferranti Ltd.), June 1976 and August 1981
Ferranti Review (Ferranti Ltd.), August 1980
FINAN, W.: 'The international transfer of semiconductor technology through US based firms'. National Bureau of Economic Research, New York, 1975
Hitachi Publication: 'Gate arrays CMOS & BiCMOS from concept to silicon' (undated)
HOGAN, C.L.: 'An address to Semicon/Europe 75', Zurich, 4th November 1975
KAHNG, D. and ATALLA, M.M.: 'Silicon–Silicon-dioxide field induced surface devices'. IRE/AIEE Solid State Device Res. Conference, Carnegie Institute of Technology, Pittsburgh PA, 1960
KELLETT, C.M., KING, W.J. and LEITH, F.A.: 'High energy implantation of materials'. Scientific Report No.1, Ion Physics Corporation, AD 651313, Jan. 1967
MACKINTOSH, I.M.: 'Profile of the worldwide semiconductor industry', 1982
MACKINTOSH, I.M.: 'Profile of the worldwide semiconductor industry', 1986
MACKINTOSH, I.M.: 'Microelectronics into the 80s', 1979
Organisation for Economic Co–operation and Development: 'Electronic components'. Gaps in Technology Series, Paris, 1968
TEAL, G.K.: 'Some recent developments in silicon and germanium materials and devices'. Presented at the National Conference on Airborne Electronics, Dayton, Ohio, 10th May 1954
'The microprocessor, the device of the future'. Texas Instruments Publication, TI (Bedford), Undated
25th anniversary observance transistor radio and silicon transistor. Texas Instruments Publication (Dallas), undated

Index

Printed in the USA
CPSIA information can be obtained
at www.ICGtesting.com
JSHW011510221024
72173JS00005B/1264

9 780863 412271